CHAPMAN & HALL/CRC
Texts in Statistical Science Series

Series Editors
Chris Chatfield, *University of Bath, UK*
Martin Tanner, *Northwestern University, USA*
Jim Zidek, *University of British Columbia, Canada*

Texts in Statistical Science

Linear Models
with R

Julian J. Faraway

CHAPMAN & HALL/CRC

A CRC Press Company

Boca Raton   London   New York   Washington, D.C.

Library of Congress Cataloging-in-Publication Data

Faraway, Julian James.
    Linear models with R / Julian J. Faraway.
        p. cm. — (Chapman & Hall/CRC texts in statistical science series ; v. 63)
    Includes bibliographical references and index.
    ISBN 1-58488-425-8 (alk. paper)
        1. Analysis of variance. 2. Regression analysis. I. Title. II. Texts in statistical science ;
    v. 63.

QA279.F37 2004
519.5'38--dc22   2004051916

Visit the CRC Press Web site at www.crcpress.com

© 2005 by Chapman & Hall/CRC

No claim to original U.S. Government works
International Standard Book Number 1-58488-425-8
Library of Congress Card Number 2004051916
Printed in the United States of America   4 5 6 7 8 9 0
Printed on acid-free paper

Contents

Preface

There are many books on regression and analysis of variance. These books expect different levels of preparedness and place different emphases on the material. This book is not introductory. It presumes some knowledge of basic statistical theory and practice. Readers are expected to know the essentials of statistical inference such as estimation, hypothesis testing and confidence intervals. A basic knowledge of data analysis is presumed. Some linear algebra and calculus are also required.

The emphasis of this text is on the practice of regression and analysis of variance. The objective is to learn what methods are available and more importantly, when they should be applied. Many examples are presented to clarify the use of the techniques and to demonstrate what conclusions can be made. There is relatively less emphasis on mathematical theory, partly because some prior knowledge is assumed and partly because the issues are better tackled elsewhere. Theory is important because it guides the approach we take. I take a wider view of statistical theory. It is not just the formal theorems. Qualitative statistical concepts are just as important in statistics because these enable us to actually do it rather than just talk about it. These qualitative principles are harder to learn because they are difficult to state precisely but they guide the successful experienced statistician.

Data analysis cannot be learned without actually doing it. This means using a statistical computing package. There is a wide choice of such packages. They are designed for different audiences and have different strengths and weaknesses. I have chosen to use R (Ref. Ihaka and Gentleman (1996) and R Development Core Team (2003)). Why have I used R? There are several reasons.

1. *Versatility*. R is also a programming language, so I am not limited by the procedures that are preprogrammed by a package. It is relatively easy to program new methods in R.

2. *Interactivity*. Data analysis is inherently interactive. Some older statistical packages were designed when computing was more expensive and batch processing of computations was the norm. Despite improvements in hardware, the old batch processing paradigm lives on in their use. R does one thing at a time, allowing us to make changes on the basis of what we see during the analysis.

3. *Freedom*. R is based on S from which the commercial package S-plus is derived. R itself is open-source software and may be obtained free of charge to all. Linux, Macintosh, Windows and other UNIX versions are maintained and can be obtained from the R-project at www.r-project.org. R is mostly compatible with S-plus, meaning that S-plus could easily be used for most of the examples provided in this book.

4. *Popularity.* SAS is the most common statistics package in general use but R or S is
   most popular with researchers in statistics. A look at common statistical journals
   confirms this popularity. R is also popular for quantitative applications in finance.

Getting Started with R

R requires some effort to learn. Such effort will be repaid with increased produc-
tivity. You can learn how to obtain R in Appendix A along with instructions on the
installation of additional software and data used in this book.

This book is not an introduction to R. Appendix B provides a brief introduction
to the language, but alone is insufficient. I have intentionally included in the text
all the commands used to produce the output seen in this book. This means that
you can reproduce these analyses and experiment with changes and variations before
fully understanding R. You may choose to start working through this text before
learning R and pick it up as you go. Free introductory guides to R may be obtained
from the R project Web site at www.r-project.org. Introductory books have
been written by Dalgaard (2002) and Maindonald and Braun (2003). Venables and
Ripley (2002) also have an introduction to R along with more advanced material. Fox
(2002) is intended as a companion to a standard regression text. You may also find
Becker, Chambers, and Wilks (1998) and Chambers and Hastie (1991) to be useful
references to the S language. Ripley and Venables (2000) wrote a more advanced text
on programming in S or R.

The Web site for this book is at www.maths.bath.ac.uk/~jjf23/LMR
where data described in this book appear. Updates and errata will appear there also.

Thanks to the builders of R without whom this book would not have been possible.

CHAPTER 1

Introduction

1.1 Before You Start

Statistics starts with a problem, proceeds with the collection of data, continues with the data analysis and finishes with conclusions. It is a common mistake of inexperienced statisticians to plunge into a complex analysis without paying attention to what the objectives are or even whether the data are appropriate for the proposed analysis. Look before you leap!

> The formulation of a problem is often more essential than its solution which may be merely a matter of mathematical or experimental skill. *Albert Einstein*

To formulate the problem correctly, you must:

1. Understand the physical background. Statisticians often work in collaboration with others and need to understand something about the subject area. Regard this as an opportunity to learn something new rather than a chore.

2. Understand the objective. Again, often you will be working with a collaborator who may not be clear about what the objectives are. Beware of "fishing expeditions" — if you look hard enough, you will almost always find something, but that something may just be a coincidence.

3. Make sure you know what the client wants. You can often do quite different analyses on the same dataset. Sometimes statisticians perform an analysis far more complicated than the client really needed. You may find that simple descriptive statistics are all that are needed.

4. Put the problem into statistical terms. This is a challenging step and where irreparable errors are sometimes made. Once the problem is translated into the language of statistics, the solution is often routine. Difficulties with this step explain why artificial intelligence techniques have yet to make much impact in application to statistics. Defining the problem is hard to program.

That a statistical method can read in and process the data is not enough. The results of an inapt analysis may be meaningless.

It is also important to understand how the data were collected.

- Are the data observational or experimental? Are the data a sample of convenience or were they obtained via a designed sample survey. How the data were collected has a crucial impact on what conclusions can be made.

- Is there nonresponse? The data you do not see may be just as important as the data you do see.

- Are there missing values? This is a common problem that is troublesome and time consuming to handle.

- How are the data coded? In particular, how are the qualitative variables represented?

- What are the units of measurement?

- Beware of data entry errors and other corruption of the data. This problem is all too common — almost a certainty in any real dataset of at least moderate size. Perform some data sanity checks.

1.2 Initial Data Analysis

This is a critical step that should always be performed. It looks simple but it is vital. You should make numerical summaries such as means, standard deviations (SDs), maximum and minimum, correlations and whatever else is appropriate to the specific dataset. Equally important are graphical summaries. There is a wide variety of techniques to choose from. For one variable at a time, you can make boxplots, histograms, density plots and more. For two variables, scatterplots are standard while for even more variables, there are numerous good ideas for display including interactive and dynamic graphics. In the plots, look for outliers, data-entry errors, skewed or unusual distributions and structure. Check whether the data are distributed according to prior expectations.

Getting data into a form suitable for analysis by cleaning out mistakes and aberrations is often time consuming. It often takes more time than the data analysis itself. In this course, all the data will be ready to analyze, but you should realize that in practice this is rarely the case.

Let's look at an example. The National Institute of Diabetes and Digestive and Kidney Diseases conducted a study on 768 adult female Pima Indians living near Phoenix. The following variables were recorded: number of times pregnant, plasma glucose concentration at 2 hours in an oral glucose tolerance test, diastolic blood pressure (mmHg), triceps skin fold thickness (mm), 2-hour serum insulin (mu U/ml), body mass index (weight in kg/(height in m^2)), diabetes pedigree function, age (years) and a test whether the patient showed signs of diabetes (coded zero if negative, one if positive). The data may be obtained from UCI Repository of machine learning databases at www.ics.uci.edu/~mlearn/MLRepository.html.

Of course, before doing anything else, one should find out the purpose of the study and more about how the data were collected. However, let's skip ahead to a look at the data:

```
> library(faraway)
> data(pima)
> pima
    pregnant glucose diastolic triceps insulin  bmi diabetes age
1          6     148        72      35       0 33.6    0.627  50
2          1      85        66      29       0 26.6    0.351  31
3          8     183        64       0       0 23.3    0.672  32
... much deleted ...
```

```
768         1       93        70       31      0 30.4     0.315   23
```

The library(faraway) command makes the data used in this book available. You need to install this package first as explained in Appendix A. We have explicitly written this command here, but in all subsequent chapters, we will assume that you have already issued this command if you plan to use data mentioned in the text. If you get an error message about data not being found, it may be that you have forgotten to type this.

The command data(pima) calls up this particular dataset. Simply typing the name of the *data frame*, pima, prints out the data. It is too long to show it all here. For a dataset of this size, one can just about visually skim over the data for anything out of place, but it is certainly easier to use summary methods.

We start with some numerical summaries:

```
> summary(pima)
    pregnant            glucose           diastolic           triceps
 Min.    : 0.00    Min.    :  0     Min.    :  0.0    Min.    : 0.0
 1st Qu.: 1.00    1st Qu.: 99     1st Qu.: 62.0    1st Qu.: 0.0
 Median : 3.00    Median :117     Median : 72.0    Median :23.0
 Mean    : 3.85    Mean    :121     Mean    : 69.1    Mean    :20.5
 3rd Qu.: 6.00    3rd Qu.:140     3rd Qu.: 80.0    3rd Qu.:32.0
 Max.    :17.00    Max.    :199     Max.    :122.0    Max.    :99.0
     insulin             bmi             diabetes             age
 Min.    :  0.0    Min.    :  0.0    Min.    :0.078    Min.    :21.0
 1st Qu.:  0.0    1st Qu.:27.3    1st Qu.:0.244    1st Qu.:24.0
 Median :  30.5    Median :32.0    Median :0.372    Median :29.0
 Mean    :  79.8    Mean    :32.0    Mean    :0.472    Mean    :33.2
 3rd Qu.:127.2    3rd Qu.:36.6    3rd Qu.:0.626    3rd Qu.:41.0
 Max.    :846.0    Max.    :67.1    Max.    :2.420    Max.    :81.0
        test
 Min.    :0.000
 1st Qu.:0.000
 Median :0.000
 Mean    :0.349
 3rd Qu.:1.000
 Max.    :1.000
```

The summary() command is a quick way to get the usual univariate summary information. At this stage, we are looking for anything unusual or unexpected, perhaps indicating a data-entry error. For this purpose, a close look at the minimum and maximum values of each variable is worthwhile. Starting with pregnant, we see a maximum value of 17. This is large, but not impossible. However, we then see that the next five variables have minimum values of zero. No blood pressure is not good for the health — something must be wrong. Let's look at the sorted values:

```
> sort(pima$diastolic)
  [1]    0   0   0   0   0   0   0   0   0   0   0   0   0   0   0
 [16]    0   0   0   0   0   0   0   0   0   0   0   0   0   0   0
 [31]    0   0   0   0   0  24  30  30  38  40  44  44  44  44  46
...etc...
```

We see that the first 35 values are zero. The description that comes with the data says nothing about it but it seems likely that the zero has been used as a missing value code. For one reason or another, the researchers did not obtain the blood pressures of 35 patients. In a real investigation, one would likely be able to question the researchers about what really happened. Nevertheless, this does illustrate the kind of misunderstanding that can easily occur. A careless statistician might overlook these presumed missing values and complete an analysis assuming that these were real observed zeros. If the error was later discovered, they might then blame the researchers for using zero as a missing value code (not a good choice since it is a valid value for some of the variables) and not mentioning it in their data description. Unfortunately such oversights are not uncommon, particularly with datasets of any size or complexity. The statistician bears some share of responsibility for spotting these mistakes.

We set all zero values of the five variables to NA which is the missing value code used by R:

```
> pima$diastolic[pima$diastolic == 0]   <- NA
> pima$glucose[pima$glucose == 0] <- NA
> pima$triceps[pima$triceps == 0]   <- NA
> pima$insulin[pima$insulin == 0] <- NA
> pima$bmi[pima$bmi == 0] <- NA
```

The variable test is not quantitative but categorical. Such variables are also called *factors*. However, because of the numerical coding, this variable has been treated as if it were quantitative. It is best to designate such variables as factors so that they are treated appropriately. Sometimes people forget this and compute stupid statistics such as the "average zip code."

```
> pima$test <- factor(pima$test)
> summary(pima$test)
  0   1
500 268
```

We now see that 500 cases were negative and 268 were positive. It is even better to use descriptive labels:

```
> levels(pima$test) <- c("negative","positive")
> summary(pima)
    pregnant            glucose            diastolic            triceps
Min.    : 0.000    Min.    : 44.0    Min.    : 24.0    Min.    :  7.00
1st Qu.: 1.000    1st Qu.: 99.0    1st Qu.: 64.0    1st Qu.: 22.00
Median : 3.000    Median :117.0    Median : 72.0    Median : 29.00
Mean    : 3.845    Mean    :121.7    Mean    : 72.4    Mean    : 29.15
3rd Qu.: 6.000    3rd Qu.:141.0    3rd Qu.: 80.0    3rd Qu.: 36.00
Max.    :17.000    Max.    :199.0    Max.    :122.0    Max.    : 99.00
                   NA's    :  5.0    NA's    : 35.0    NA's    :227.00
    insulin             bmi              diabetes             age
Min.    : 14.00    Min.    :18.20    Min.    :0.0780    Min.    :21.00
1st Qu.: 76.25    1st Qu.:27.50    1st Qu.:0.2437    1st Qu.:24.00
Median :125.00    Median :32.30    Median :0.3725    Median :29.00
Mean    :155.55    Mean    :32.46    Mean    :0.4719    Mean    :33.24
3rd Qu.:190.00    3rd Qu.:36.60    3rd Qu.:0.6262    3rd Qu.:41.00
```

```
Max.    :846.00    Max.    :67.10    Max.    :2.4200    Max.    :81.00
NA's    :374.00    NA's    :11.00
           test
negative:500
positive:268
```

Now that we have cleared up the missing values and coded the data appropriately, we are ready to do some plots. Perhaps the most well-known univariate plot is the histogram:

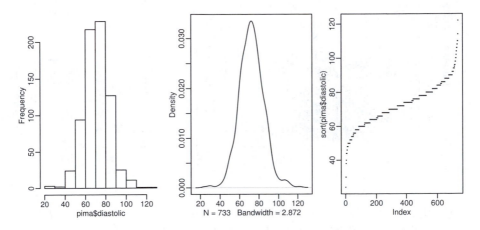

Figure 1.1 *The first panel shows a histogram of the diastolic blood pressures, the second shows a kernel density estimate of the same, while the third shows an index plot of the sorted values.*

```
> hist(pima$diastolic)
```

as seen in the first panel of Figure 1.1. We see a bell-shaped distribution for the diastolic blood pressures centered around 70. The construction of a histogram requires the specification of the number of bins and their spacing on the horizontal axis. Some choices can lead to histograms that obscure some features of the data. R specifies the number and spacing of bins given the size and distribution of the data, but this choice is not foolproof and misleading histograms are possible. For this reason, some prefer to use kernel density estimates, which are essentially a smoothed version of the histogram (see Simonoff (1996) for a discussion of the relative merits of histograms and kernel estimates):

```
> plot(density(pima$diastolic,na.rm=TRUE))
```

The kernel estimate may be seen in the second panel of Figure 1.1. We see that this plot avoids the distracting blockiness of the histogram. Another alternative is to simply plot the sorted data against its index:

```
> plot(sort(pima$diastolic),pch=".")
```

The advantage of this is that we can see all the cases individually. We can see the distribution and possible outliers. We can also see the discreteness in the measurement of blood pressure — values are rounded to the nearest even number and hence we see the "steps" in the plot.

Now note a couple of bivariate plots, as seen in Figure 1.2:

```
> plot(diabetes ~ diastolic,pima)
> plot(diabetes ~ test,pima)
```

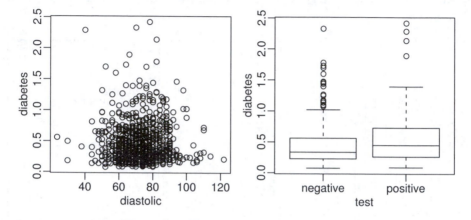

Figure 1.2 *The first panel shows scatterplot of the diastolic blood pressures against diabetes function and the second shows boxplots of diastolic blood pressure broken down by test result.*

First, we see the standard scatterplot showing two quantitative variables. Second, we see a side-by-side boxplot suitable for showing a quantitative and a qualititative variable. Also useful is a scatterplot matrix, not shown here, produced by:

```
> pairs(pima)
```

We will be seeing more advanced plots later, but the numerical and graphical summaries presented here are sufficient for a first look at the data.

1.3 When to Use Regression Analysis

Regression analysis is used for explaining or modeling the relationship between a single variable Y, called the *response, output* or *dependent* variable; and one or more *predictor, input, independent* or *explanatory* variables, $X_1, \ldots, X_p$. When $p = 1$, it is called *simple* regression but when $p > 1$ it is called *multiple* regression or sometimes *multivariate* regression. When there is more than one Y, then it is called *multivariate multiple* regression which we will not be covering explicity here, although you can just do separate regressions on each Y.

The response must be a continuous variable, but the explanatory variables can be continuous, discrete or categorical, although we leave the handling of categorical explanatory variables to later in the book. Taking the example presented above, a regression with `diastolic` and `bmi` as X's and `diabetes` as Y would be a multiple regression involving only quantitative variables which we shall be tackling shortly. A regression with `diastolic` and `test` as X's and `bmi` as Y would have one predictor that is quantitative and one that is qualitative, which we will consider later in Chapter 13 on *analysis of covariance*. A regression with `test` as X and `diastolic` as Y involves just qualitative predictors — a topic called *analysis of variance (ANOVA)*, although this would just be a simple two sample situation. A regression of `test` as Y on `diastolic` and `bmi` as predictors would involve a qualitative response. A *logistic regression* could be used, but this will not be covered in this book.

Regression analyses have several possible objectives including:

1. Prediction of future observations

2. Assessment of the effect of, or relationship between, explanatory variables and the response

3. A general description of data structure

Extensions exist to handle multivariate responses, binary responses (logistic regression analysis) and count responses (Poisson regression) among others.

1.4 History

Regression-type problems were first considered in the 18th century to aid navigation with the use of astronomy. Legendre developed the method of least squares in 1805. Gauss claimed to have developed the method a few years earlier and in 1809 showed that least squares is the optimal solution when the errors are normally distributed. The methodology was used almost exclusively in the physical sciences until later in the 19th century. Francis Galton coined the term *regression to mediocrity* in 1875 in reference to the simple regression equation in the form:

$$\frac{y - \bar{y}}{SD_y} = r\frac{(x - \bar{x})}{SD_x}$$

where r is the correlation between x and y. Galton used this equation to explain the phenomenon that sons of tall fathers tend to be tall but not as tall as their fathers, while sons of short fathers tend to be short but not as short as their fathers. This phenomenom is called the *regression effect*. See Stigler (1986) for more of the history.

We can illustrate this effect with some data on scores from a course taught using this book. In Figure 1.3, we see a plot of midterm against final scores. We scale each variable to have mean zero and SD one so that we are not distracted by the relative difficulty of each exam and the total number of points possible. Furthermore, this simplifies the regression equation to:

$$y = rx$$

```
> data(stat500)
> stat500 <- data.frame(scale(stat500))
> plot(final ~ midterm, stat500)
> abline(0,1)
```

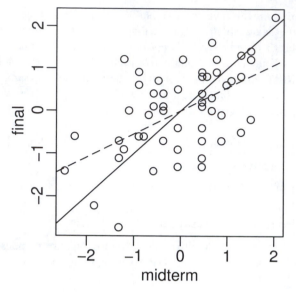

Figure 1.3 *The final and midterm scores in standard units. The least squares fit is shown with a dotted line while y = x is shown with a solid line.*

We have added the $y = x$ (solid) line to the plot. Now a student scoring, say 1 SD above average on the midterm might reasonably expect to do equally well on the final. We compute the least squares regression fit and plot the regression line (more on the details later). We also compute the correlations:

```
> g <- lm(final ~ midterm,stat500)
> abline(coef(g),lty=5)
> cor(stat500)
           midterm      final        hw     total
midterm  1.00000  0.545228  0.272058  0.84446
final    0.54523  1.000000  0.087338  0.77886
hw       0.27206  0.087338  1.000000  0.56443
total    0.84446  0.778863  0.564429  1.00000
```

The regression fit is the dotted line in Figure 1.3 and is always shallower than the $y = x$ line. We see that a student scoring 1 SD above average on the midterm is predicted to score only 0.545 SDs above average on the final

Correspondingly, a student scoring below average on the midterm might expect to do relatively better in the final, although still below average.

If exams managed to measure the ability of students perfectly, then provided that ability remained unchanged from midterm to final, we would expect to see an exact correlation. Of course, it is too much to expect such a perfect exam and some variation is inevitably present. Furthermore, individual effort is not constant. Getting a high score on the midterm can partly be attributed to skill, but also a certain amount of luck. One cannot rely on this luck to be maintained in the final. Hence we see the "regression to mediocrity."

Of course this applies to any (x,y) situation like this — an example is the so-called sophomore jinx in sports when a new star has a so-so second season after a great first year. Although in the father–son example, it does predict that successive descendants will come closer to the mean; it does not imply the same of the population in general since random fluctuations will maintain the variation. In many other applications of regression, the regression effect is not of interest, so it is unfortunate that we are now stuck with this rather misleading name.

Regression methodology developed rapidly with the advent of high-speed computing. Just fitting a regression model used to require extensive hand calculation. As computing hardware has improved, the scope for analysis has widened.

Exercises

1. The dataset teengamb concerns a study of teenage gambling in Britain. Make a numerical and graphical summary of the data, commenting on any features that you find interesting. Limit the output you present to a quantity that a busy reader would find sufficient to get a basic understanding of the data.

2. The dataset uswages is drawn as a sample from the Current Population Survey in 1988. Make a numerical and graphical summary of the data as in the previous question.

3. The dataset prostate is from a study on 97 men with prostate cancer who were due to receive a radical prostatectomy. Make a numerical and graphical summary of the data as in the first question.

4. The dataset sat comes from a study entitled "Getting What You Pay For: The Debate Over Equity in Public School Expenditures." Make a numerical and graphical summary of the data as in the first question.

5. The dataset divusa contains data on divorces in the United States from 1920 to 1996. Make a numerical and graphical summary of the data as in the first question.

CHAPTER 2

Estimation

2.1 Linear Model

Suppose we want to model the response Y in terms of three predictors, X_1, X_2 and X_3. One very general form for the model would be:

$$Y = f(X_1, X_2, X_3) + \varepsilon$$

where f is some unknown function and ε is the error in this representation. ε is additive in this instance, but could enter in some more general form. Still, if we assume that f is a smooth, continuous function, that still leaves a very wide range of possibilities. Even with just three predictors, we typically will not have enough data to try to estimate f directly. So we usually have to assume that it has some more restricted form, perhaps linear as in:

$$Y = \beta_0 + \beta_1 X_1 + \beta_2 X_2 + \beta_3 X_3 + \varepsilon$$

where β_i, $i = 0, 1, 2, 3$ are unknown *parameters*. β_0 is called the *intercept* term. Thus the problem is reduced to the estimation of four parameters rather than the infinite dimensional f. In a linear model the *parameters enter linearly* — the predictors themselves do not have to be linear. For example:

$$Y = \beta_0 + \beta_1 X_1 + \beta_2 \log X_2 + \beta_3 X_1 X_2 + \varepsilon$$

is a linear model, but:

$$Y = \beta_0 + \beta_1 X_1^{\beta_2} + \varepsilon$$

is not. Some relationships can be transformed to linearity — for example, $y = \beta_0 x_1^\beta \varepsilon$ can be linearized by taking logs. Linear models seem rather restrictive, but because the predictors can be transformed and combined in any way, they are actually very flexible. The term linear is often used in everyday speech as almost a synonym for simplicity. This gives the casual observer the impression that linear models can only handle small simple datasets. This is far from the truth — linear models can easily be expanded and modified to handle complex datasets. Linear is also used to refer to straight lines, but linear models can be curved. Truly nonlinear models are rarely absolutely necessary and most often arise from a theory about the relationships between the variables, rather than an empirical investigation.

2.2 Matrix Representation

If we have a response Y and three predictors, X_1, X_2 and X_3, the data might be presented in tabular form like this:

$$
\begin{array}{cccc}
y_1 & x_{11} & x_{12} & x_{13} \\
y_2 & x_{21} & x_{22} & x_{23} \\
\cdots & & \cdots & \\
y_n & x_{n1} & x_{n2} & x_{n3}
\end{array}
$$

where n is the number of observations, or *cases*, in the dataset.

Given the actual data values, we may write the model as:

$$
y_i = \beta_0 + \beta_1 x_{i1} + \beta_2 x_{i2} + \beta_3 x_{i3} + \varepsilon_i \quad i = 1, \ldots, n
$$

but the use of subscripts becomes inconvenient and conceptually obscure. We will find it simpler both notationally and theoretically to use a matrix/vector representation. The regression equation is written as:

$$
y = X\beta + \varepsilon
$$

where $y = (y_1, \ldots, y_n)^T$, $\varepsilon = (\varepsilon_1, \ldots, \varepsilon_n)^T$, $\beta = (\beta_0, \ldots, \beta_3)^T$ and:

$$
X = \begin{pmatrix}
1 & x_{11} & x_{12} & x_{13} \\
1 & x_{21} & x_{22} & x_{23} \\
\cdots & & \cdots & \\
1 & x_{n1} & x_{n2} & x_{n3}
\end{pmatrix}
$$

The column of ones incorporates the intercept term. One simple example is the *null model* where there is no predictor and just a mean $y = \mu + \varepsilon$:

$$
\begin{pmatrix} y_1 \\ \cdots \\ y_n \end{pmatrix} = \begin{pmatrix} 1 \\ \cdots \\ 1 \end{pmatrix} \mu + \begin{pmatrix} \varepsilon_1 \\ \cdots \\ \varepsilon_n \end{pmatrix}
$$

We can assume that $E\varepsilon = 0$ since if this were not so, we could simply absorb the nonzero expectation for the error into the mean μ to get a zero expectation.

2.3 Estimating β

The regression model, $y = X\beta + \varepsilon$, partitions the response into a systematic component $X\beta$ and a random component ε. We would like to choose β so that the systematic part explains as much of the response as possible. Geometrically speaking, the response lies in an n-dimensional space, that is, $y \in \mathbb{R}^n$ while $\beta \in \mathbb{R}^p$ where p is the number of parameters. If we include the intercept then p is the number of predictors plus one. We will use this definition of p from now on. It is easy to get confused as to whether p is the number of predictors or parameters, as different authors use different conventions, so be careful.

The problem is to find β so that $X\beta$ is as close to Y as possible. The best choice, the estimate $\hat{\beta}$, is apparent in the geometrical representation seen in Figure 2.1. $\hat{\beta}$ is, in this sense, the best estimate of β within the model space. The response

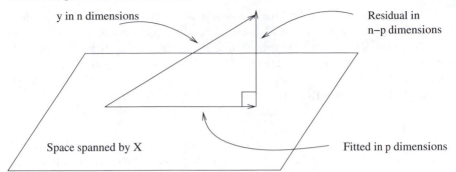

Figure 2.1 *Geometrical representation of the estimation β. The data vector Y is projected orthogonally onto the model space spanned by X. The fit is represented by projection $\hat{y} = X\hat{\beta}$ with the difference between the fit and the data represented by the residual vector ê.*

predicted by the model is $\hat{y} = X\hat{\beta}$ or Hy where H is an orthogonal projection matrix. The difference between the actual response and the predicted response is denoted by ê and is called the *residuals*.

The conceptual purpose of the model is to represent, as accurately as possible, something complex, y, which is n-dimensional, in terms of something much simpler, the model, which is p-dimensional. Thus if our model is successful, the structure in the data should be captured in those p dimensions, leaving just random variation in the residuals which lie in an $(n-p)$-dimensional space. We have:

$$\text{Data} \;=\; \text{Systematic Structure} + \text{Random Variation}$$
$$n \text{ dimensions} \;=\; p \text{ dimensions} + (n-p) \text{ dimensions}$$

2.4 Least Squares Estimation

The estimation of β can also be considered from a nongeometrical point of view. We might define the best estimate of β as the one which minimizes the sum of the squared errors. The *least squares* estimate of β, called $\hat{\beta}$ minimizes:

$$\sum \varepsilon_i^2 = \varepsilon^T \varepsilon = (y - X\beta)^T (y - X\beta)$$

Differentiating with respect to β and setting to zero, we find that $\hat{\beta}$ satisfies:

$$X^T X \hat{\beta} = X^T y$$

These are called the *normal equations*. We can derive the same result using the geometrical approach. Now provided $X^T X$ is invertible:

$$
\begin{aligned}
\hat{\beta} &= (X^T X)^{-1} X^T y \\
X\hat{\beta} &= X(X^T X)^{-1} X^T y \\
\hat{y} &= Hy
\end{aligned}
$$

$H = X(X^TX)^{-1}X^T$ is called the *hat-matrix* and is the orthogonal projection of y onto the space spanned by X. H is useful for theoretical manipulations, but you usually do not want to compute it explicitly, as it is an $n \times n$ matrix which could be uncomfortably large for some datasets. The following useful quantities can now be represented using H.

The predicted or fitted values are $\hat{y} = Hy = X\hat{\beta}$ while the residuals are $\hat{\varepsilon} = y - X\hat{\beta} = y - \hat{y} = (I - H)y$. The residual sum of squares (RSS) is $\hat{\varepsilon}^T\hat{\varepsilon} = y^T(I - H)^T(I - H)y = y^T(I - H)y$.

Later, we will show that the least squares estimate is the best possible estimate of β when the errors ε are uncorrelated and have equal variance or more briefly put $\text{var } \varepsilon = \sigma^2 I$.

$\hat{\beta}$ is unbiased and has variance $(X^TX)^{-1}\sigma^2$ provided $\text{var } \varepsilon = \sigma^2 I$. Since $\hat{\beta}$ is a vector, its variance is a matrix.

We also need to estimate σ^2. We find that $E\hat{\varepsilon}^T\hat{\varepsilon} = \sigma^2(n - p)$, which suggests the estimator:

$$\hat{\sigma}^2 = \frac{\hat{\varepsilon}^T\hat{\varepsilon}}{n - p} = \frac{\text{RSS}}{n - p}$$

as an unbiased estimate of σ^2. $n - p$ is the *degrees of freedom* of the model. Sometimes you need the standard error for a particular component of $\hat{\beta}$ which can be picked out as $se(\hat{\beta}_{i-1}) = \sqrt{(X^TX)_{ii}^{-1}}\hat{\sigma}$.

2.5 Examples of Calculating $\hat{\beta}$

In a few simple models, it is possible to derive explicit formulae for $\hat{\beta}$:

1. When $y = \mu + \varepsilon$, $X = 1$ and $\beta = \mu$ so $X^TX = 1^T1 = n$ so:

$$\hat{\beta} = (X^TX)^{-1}X^Ty = \frac{1}{n}1^Ty = \bar{y}$$

2. Simple linear regression (one predictor):

$$y_i = \beta_0 + \beta_1 x_i + \varepsilon_i$$

$$\begin{pmatrix} y_1 \\ \dots \\ y_n \end{pmatrix} = \begin{pmatrix} 1 & x_1 \\ \dots & \\ 1 & x_n \end{pmatrix} \begin{pmatrix} \beta_0 \\ \beta_1 \end{pmatrix} + \begin{pmatrix} \varepsilon_1 \\ \dots \\ \varepsilon_n \end{pmatrix}$$

We can now apply the formula but a simpler approach is to rewrite the equation as:

$$y_i = \overbrace{\beta_0 + \beta_1\bar{x}}^{\beta_0'} + \beta_1(x_i - \bar{x}) + \varepsilon_i$$

so now:

$$X = \begin{pmatrix} 1 & x_1 - \bar{x} \\ \dots & \\ 1 & x_n - \bar{x} \end{pmatrix} \qquad X^TX = \begin{pmatrix} n & 0 \\ 0 & \sum_{i=1}^n (x_i - \bar{x})^2 \end{pmatrix}$$

Next work through the rest of the calculation to reconstruct the familiar estimates, that is:

$$\hat{\beta}_1 = \frac{\sum (x_i - \bar{x}) y_i}{\sum (x_i - \bar{x})^2}$$

In higher dimensions, it is usually not possible to find such explicit formulae for the parameter estimates unless $X^T X$ happens to be a simple form. So typically we need computers to fit such models. Regression has a long history, so in the time before computers became readily available, fitting even quite simple models was a tedious time consuming task. When computing was expensive, data analysis was limited. It was designed to keep calculations to a minimum and restrict the number of plots. This mindset remained in statistical practice for some time even after computing became widely and cheaply available. Now it is a simple matter to fit a multitude of models and make more plots than one could reasonably study. The challenge for the analyst is to choose among these intelligently to extract the crucial information in the data.

2.6 Gauss–Markov Theorem

$\hat{\beta}$ is a plausible estimator, but there are alternatives. Nonetheless, there are three good reasons to use least squares:

1. It results from an orthogonal projection onto the model space. It makes sense geometrically.

2. If the errors are independent and identically normally distributed, it is the maximum likelihood estimator. Loosely put, the maximum likelihood estimate is the value of β that maximizes the probability of the data that was observed.

3. The Gauss–Markov theorem states that $\hat{\beta}$ is the best linear unbiased estimate (BLUE).

To understand the Gauss–Markov theorem we first need to understand the concept of an *estimable function*. A linear combination of the parameters $\psi = c^T \beta$ is estimable if and only if there exists a linear combination $a^T y$ such that:

$$E a^T y = c^T \beta \qquad \forall \beta$$

Estimable functions include predictions of future observations, which explains why they are well worth considering. If X is of full rank, then all linear combinations are estimable.

Suppose $E \varepsilon = 0$ and var $\varepsilon = \sigma^2 I$. Suppose also that the structural part of the model, $EY = X \beta$ is correct. (Clearly these are big assumptions and so we will address the implications of this later.) Let $\psi = c^T \beta$ be an estimable function; then the Gauss–Markov theorem states that in the class of all unbiased linear estimates of ψ, $\hat{\psi} = c^T \hat{\beta}$ has the minimum variance and is unique.

We prove this theorem. Suppose $a^T y$ is some unbiased estimate of $c^T \beta$ so that:

$$\begin{aligned} E a^T y &= c^T \beta & \forall \beta \\ a^T X \beta &= c^T \beta & \forall \beta \end{aligned}$$

which means that $a^T X = c^T$. This implies that c must be in the range space of X^T which in turn implies that c is also in the range space of $X^T X$ which means there exists a λ such that $c = X^T X \lambda$ so:

$$c^T \hat{\beta} = \lambda^T X^T X \hat{\beta} = \lambda^T X^T y$$

Now we can show that the least squares estimator has the minimum variance — pick an arbitrary estimate $a^T y$ and compute its variance:

$$
\begin{aligned}
\text{var}\,(a^T y) &= \text{var}\,(a^T y - c^T \hat{\beta} + c^T \hat{\beta}) \\
&= \text{var}\,(a^T y - \lambda^T X^T y + c^T \hat{\beta}) \\
&= \text{var}\,(a^T y - \lambda^T X^T y) + \text{var}\,(c^T \hat{\beta}) + 2\text{cov}(a^T y - \lambda^T X^T y, \lambda^T X^T y)
\end{aligned}
$$

but

$$
\begin{aligned}
\text{cov}(a^T y - \lambda^T X^T y, \lambda^T X^T y) &= (a^T - \lambda^T X^T)\sigma^2 I X \lambda \\
&= (a^T X - \lambda^T X^T X)\sigma^2 I \lambda \\
&= (c^T - c^T)\sigma^2 I \lambda = 0
\end{aligned}
$$

so

$$\text{var}\,(a^T y) = \text{var}\,(a^T y - \lambda^T X^T y) + \text{var}\,(c^T \hat{\beta})$$

Now since variances cannot be negative, we see that:

$$\text{var}\,(a^T y) \geq \text{var}\,(c^T \hat{\beta})$$

In other words, $c^T \hat{\beta}$ has minimum variance. It now remains to show that it is unique. There will be equality in the above relationship if $\text{var}\,(a^T y - \lambda^T X^T y) = 0$ which would require that $a^T - \lambda^T X^T = 0$ which means that $a^T y = \lambda^T X^T y = c^T \hat{\beta}$. So equality occurs only if $a^T y = c^T \hat{\beta}$ so the estimator is unique. This completes the proof.

The Gauss–Markov theorem shows that the least squares estimate $\hat{\beta}$ is a good choice, but it does require that the errors are uncorrelated and have equal variance. Even if the errors behave, but are nonnormal, then nonlinear or biased estimates may work better. So this theorem does not tell one to use least squares all the time; it just strongly suggests it unless there is some strong reason to do otherwise. Situations where estimators other than ordinary least squares should be considered are:

1. When the errors are correlated or have unequal variance, generalized least squares should be used. See Section 6.1.

2. When the error distribution is long-tailed, then robust estimates might be used. Robust estimates are typically not linear in y. See Section 6.4.

3. When the predictors are highly correlated (collinear), then biased estimators such as ridge regression might be preferable. See Chapter 9.

2.7 Goodness of Fit

It is useful to have some measure of how well the model fits the data. One common choice is R^2, the so-called *coefficient of determination* or *percentage of variance explained*:

$$R^2 = 1 - \frac{\sum(\hat{y}_i - y_i)^2}{\sum(y_i - \bar{y})^2} = 1 - \frac{\text{RSS}}{\text{Total SS(Corrected for Mean)}}$$

Its range is $0 \le R^2 \le 1$ — values closer to 1 indicating better fits. For simple linear regression $R^2 = r^2$ where r is the correlation between x and y. An equivalent definition is:

$$R^2 = \frac{\sum(\hat{y}_i - \bar{y})^2}{\sum(y_i - \bar{y})^2}$$

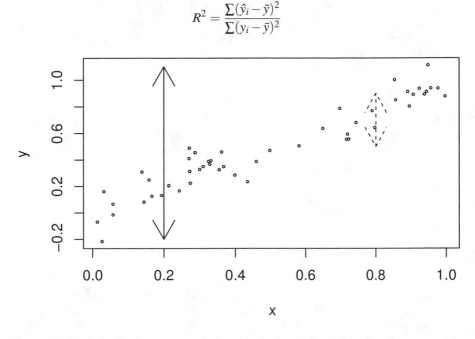

Figure 2.2 *Variation in the response y when x is known is denoted by dotted arrows while variation in y when x is unknown is shown with the solid arrows.*

The graphical intuition behind R^2 is seen in Figure 2.2. Suppose you want to predict y. If you do not know x, then your best prediction is $\bar{y}$, but the variability in this prediction is high. If you do know x, then your prediction will be given by the regression fit. This prediction will be less variable provided there is some relationship between x and y. R^2 is one minus the ratio of the sum of squares for these two predictions. Thus for perfect predictions the ratio will be zero and R^2 will be one.

R^2 as defined here does not make any sense if you do not have an intercept in your model. This is because the denominator in the definition of R^2 has a null model with an intercept in mind when the sum of squares is calculated. Alternative definitions of R^2 are possible when there is no intercept, but the same graphical intuition is not available and the R^2s obtained in this way should not be compared to those for models with an intercept. Beware of high R^2s reported from models without an intercept.

What is a good value of R^2? It depends on the area of application. In the biological and social sciences, variables tend to be more weakly correlated and there is a lot of noise. We would expect lower values for R^2 in these areas — a value of 0.6 might be considered good. In physics and engineering, where most data come from closely controlled experiments, we expect to get much higher R^2s and a value of 0.6 would be considered low. Of course, I generalize excessively here so some experience with the particular area is necessary for you to judge your R^2s well.

An alternative measure of fit is $\hat{\sigma}$. This quantity is directly related to the standard errors of estimates of β and predictions. The advantage is that $\hat{\sigma}$ is measured in the units of the response and so may be directly interpreted in the context of the particular dataset. This may also be a disadvantage in that one must understand whether the practical significance of this measure whereas R^2, being unitless, is easy to understand.

2.8 Example

Now let's look at an example concerning the number of species of tortoise on the various Galápagos Islands. There are 30 cases (Islands) and seven variables in the dataset. We start by reading the data into R and examining it (remember you first need to load the book data with the library(faraway) command):

```
> data(gala)
> gala
```

	Species	Endemics	Area	Elevation	Nearest	Scruz
Baltra	58	23	25.09	346	0.6	0.6
Bartolome	31	21	1.24	109	0.6	26.3

...

The variables are Species — the number of species of tortoise found on the island, Endemics — the number of endemic species, Area — the area of the island (km^2), Elevation — the highest elevation of the island (m), Nearest — the distance from the nearest island (km), Scruz — the distance from Santa Cruz Island (km), Adjacent — the area of the adjacent island (km^2).

The data were presented by Johnson and Raven (1973) and also appear in Weisberg (1985). I have filled in some missing values for simplicity (see Chapter 12 for how this can be done). Fitting a linear model in R is done using the lm() command. Notice the syntax for specifying the predictors in the model. This is part of the *Wilkinson–Rogers* notation. In this case, since all the variables are in the gala data frame, we must use the data= argument:

```
> mdl <- lm(Species ~ Area + Elevation + Nearest + Scruz
      + Adjacent, data=gala)
> summary(mdl)
Call:
lm(formula = Species ~ Area + Elevation + Nearest + Scruz
    + Adjacent, data = gala)

Residuals:
    Min      1Q   Median      3Q      Max
-111.68   -34.90   -7.86    33.46   182.58
```

EXAMPLE   19

```
Coefficients:
            Estimate Std. Error t value Pr(>|t|)
(Intercept)  7.06822   19.15420    0.37   0.7154
Area        -0.02394    0.02242   -1.07   0.2963
Elevation    0.31946    0.05366    5.95  3.8e-06
Nearest      0.00914    1.05414    0.01   0.9932
Scruz       -0.24052    0.21540   -1.12   0.2752
Adjacent    -0.07480    0.01770   -4.23   0.0003

Residual standard error: 61 on 24 degrees of freedom
Multiple R-Squared: 0.766,        Adjusted R-squared: 0.717
F-statistic: 15.7 on 5 and 24 DF,  p-value: 6.84e-07
```

We can identify several useful quantities in this output. Other statistical packages tend to produce output quite similar to this. One useful feature of R is that it is possible to directly calculate quantities of interest. Of course, it is not necessary here because the lm() function does the job, but it is very useful when the statistic you want is not part of the prepackaged functions.

First, we make the X-matrix:

```
> x <- model.matrix( ~ Area + Elevation + Nearest + Scruz
 + Adjacent,gala)
```

and here is the response y:

```
> y <- gala$Species
```

Now let's construct $(X^TX)^{-1}$. t() does transpose and %*% does matrix multiplication. solve(A) computes A^{-1} while solve(A,b) solves $Ax = b$:

```
> xtxi <- solve(t(x) %*% x)
```

We can get $\hat{\beta}$ directly, using $(X^TX)^{-1}X^Ty$:

```
> xtxi %*% t(x) %*% y
               [,1]
1          7.068221
Area      -0.023938
Elevation  0.319465
Nearest    0.009144
Scruz     -0.240524
Adjacent  -0.074805
```

This is a very bad way to compute $\hat{\beta}$. It is inefficient and can be very inaccurate when the predictors are strongly correlated. Such problems are exacerbated by large datasets. A better, but not perfect, way is:

```
> solve(crossprod(x,x),crossprod(x,y))
               [,1]
1          7.068221
Area      -0.023938
Elevation  0.319465
Nearest    0.009144
Scruz     -0.240524
Adjacent  -0.074805
```

where `crossprod(x,y)` computes $x^T y$. Here we get the same result as `lm()` because the data are well-behaved. In the long run, you are advised to use carefully programmed code such as found in `lm()`. To see the full details, consult a text such as Thisted (1988).

We can extract the regression quantities we need from the model object. Commonly used are `residuals()`, `fitted()`, `deviance()` which gives the RSS, `df.residual()` which gives the degrees of freedom and `coef()` which gives the $\hat{\beta}$. You can also extract other needed quantities by examining the model object and its summary:

```
> names(mdl)
 [1] "coefficients"  "residuals"      "effects"
 [4] "rank"          "fitted.values" "assign"
 [7] "qr"            "df.residual"   "xlevels"
[10] "call"          "terms"         "model"
> mdls <- summary(mdl)
> names(mdls)
 [1] "call"          "terms"          "residuals"
 [4] "coefficients"  "aliased"        "sigma"
 [7] "df"            "r.squared"      "adj.r.squared"
[10] "fstatistic"    "cov.unscaled"
```

We can estimate σ using the formula in the text above or extract it from the summary object:

```
> sqrt(deviance(mdl)/df.residual(mdl))
[1] 60.975
> mdls$sigma
[1] 60.975
```

We can also extract $(X^T X)^{-1}$ and use it to compute the standard errors for the coefficients. (`diag()` returns the diagonal of a matrix):

```
> xtxi <- mdls$cov.unscaled
> sqrt(diag(xtxi))*60.975
(Intercept)        Area   Elevation      Nearest       Scruz
  19.154139    0.022422    0.053663     1.054133    0.215402
   Adjacent
   0.017700
```

or get them from the summary object:

```
>  mdls$coef[,2]
(Intercept)        Area   Elevation      Nearest       Scruz
  19.154198    0.022422    0.053663     1.054136    0.215402
   Adjacent
   0.017700
```

Finally, we may compute or extract R^2:

```
> 1-deviance(mdl)/sum((y-mean(y))^2)
[1] 0.76585
> mdls$r.squared
[1] 0.76585
```

2.9 Identifiability

The least squares estimate is the solution to the normal equations:

$$X^T X \hat{\beta} = X^T y$$

where X is an $n \times p$ matrix. If $X^T X$ is singular and cannot be inverted, then there will be infinitely many solutions to the normal equations and $\hat{\beta}$ is at least partially unidentifiable. Unidentifiability will occur when X is not of full rank — when its columns are linearly dependent. With observational data, unidentifiability is usually caused by some oversight. Here are some examples:

1. A person's weight is measured both in pounds and kilos and both variables are entered into the model.

2. For each individual we record the number of years of preuniversity education, the number of years of university education and also the total number of years of education and put all three variables into the model.

3. We have more variables than cases, that is, $p > n$. When $p = n$, we may perhaps estimate all the parameters, but with no degrees of freedom left to estimate any standard errors or do any testing. Such a model is called *saturated.* When $p > n$, then the model is called *supersaturated.* Oddly enough, such models are considered in large-scale screening experiments used in product design and manufacture, but there is no hope of uniquely estimating all the parameters in such a model.

Such problems can be avoided by paying attention. Identifiability is more of an issue in designed experiments. Consider a simple two-sample experiment, where the treatment observations are $y_1, \ldots, y_n$ and the controls are $y_{n+1}, \ldots, y_{m+n}$. Suppose we try to model the response by an overall mean μ and group effects α_1 and α_2:

$$y_j = \mu + \alpha_i + \varepsilon_j \qquad i = 1, 2 \quad j = 1, \ldots, m+n$$

$$\begin{pmatrix} y_1 \\ \cdots \\ y_n \\ y_{n+1} \\ \cdots \\ y_{m+n} \end{pmatrix} = \begin{pmatrix} 1 & 1 & 0 \\ & \cdots & \\ 1 & 1 & 0 \\ 1 & 0 & 1 \\ \cdot & \cdot & \cdot \\ 1 & 0 & 1 \end{pmatrix} \begin{pmatrix} \mu \\ \alpha_1 \\ \alpha_2 \end{pmatrix} + \begin{pmatrix} \varepsilon_1 \\ \cdots \\ \cdots \\ \cdots \\ \varepsilon_{m+n} \end{pmatrix}$$

Now although X has three columns, it has only rank 2 — $(\mu, \alpha_1, \alpha_2)$ are not identifiable and the normal equations have infinitely many solutions. We can solve this problem by imposing some constraints, $\mu = 0$ or $\alpha_1 + \alpha_2 = 0$, for example.

Statistics packages handle nonidentifiability differently. In the regression case above, some may return error messages and some may fit models because rounding error may remove the exact identifiability. In other cases, constraints may be applied but these may be different from what you expect. By default, R fits the largest identifiable model by removing variables in the reverse order of appearance in the model formula.

Here is an example. Suppose we create a new variable for the Galápagos dataset — the difference in area between the island and its nearest neighbor:

```
> gala$Adiff <- gala$Area -gala$Adjacent
```

and add that to the model:

```
> g <- lm(Species ~ Area+Elevation+Nearest+Scruz+Adjacent
  +Adiff,gala)
> summary(g)
Coefficients: (1 not defined because of singularities)
            Estimate Std. Error t value Pr(>|t|)
(Intercept)  7.06822   19.15420    0.37   0.7154
Area        -0.02394    0.02242   -1.07   0.2963
Elevation    0.31946    0.05366    5.95 3.8e-06
Nearest      0.00914    1.05414    0.01   0.9932
Scruz       -0.24052    0.21540   -1.12   0.2752
Adjacent    -0.07480    0.01770   -4.23   0.0003
Adiff             NA         NA      NA       NA

Residual standard error: 61 on 24 degrees of freedom
Multiple R-Squared: 0.766,       Adjusted R-squared: 0.717
F-statistic: 15.7 on 5 and 24 DF,   p-value: 6.84e-07
```

We get a message about one undefined coefficient because the rank of the design matrix X is six, which is less than its seven columns. In most cases, the cause of identifiability can be revealed with some thought about the variables, but, failing that, an eigendecomposition of $X^T X$ will reveal the linear combination(s) that gave rise to the unidentifiability.

Lack of identifiability is obviously a problem, but it is usually easy to identify and work around. More problematic are cases where we are close to unidentifiability. To demonstrate this, suppose we add a small random perturbation to the third decimal place of Adiff by adding a random variate from $U[-0.005, 0.005]$ where U denotes the uniform distribution:

```
> Adiffe <- gala$Adiff+0.001*(runif(30)-0.5)
```

and now refit the model:

```
> g <- lm(Species ~ Area+Elevation+Nearest+Scruz
  +Adjacent+Adiffe,gala)
> summary(g)
Coefficients:
            Estimate Std. Error t value Pr(>|t|)
(Intercept) 7.14e+00   1.95e+01    0.37     0.72
Area       -2.38e+04   4.70e+04   -0.51     0.62
Elevation   3.12e-01   5.67e-02    5.50 1.4e-05
Nearest     1.38e-01   1.10e+00    0.13     0.90
Scruz      -2.50e-01   2.20e-01   -1.14     0.27
Adjacent    2.38e+04   4.70e+04    0.51     0.62
Adiffe      2.38e+04   4.70e+04    0.51     0.62

Residual standard error: 61.9 on 23 degrees of freedom
Multiple R-Squared: 0.768,       Adjusted R-squared: 0.708
F-statistic: 12.7 on 6 and 23 DF,   p-value: 2.58e-06
```

Notice that now all parameters are estimated, but the standard errors are very large because we cannot estimate them in a stable way. We deliberately caused this problem so we know the cause but in general we need to be able to identify such situations. We do this in Section 5.3.

Exercises

1. The dataset `teengamb` concerns a study of teenage gambling in Britain. Fit a regression model with the expenditure on gambling as the response and the sex, status, income and verbal score as predictors. Present the output.

   (a) What percentage of variation in the response is explained by these predictors?
   (b) Which observation has the largest (positive) residual? Give the case number.
   (c) Compute the mean and median of the residuals.
   (d) Compute the correlation of the residuals with the fitted values.
   (e) Compute the correlation of the residuals with the income.
   (f) For all other predictors held constant, what would be the difference in predicted expenditure on gambling for a male compared to a female?

2. The dataset `uswages` is drawn as a sample from the Current Population Survey in 1988. Fit a model with weekly wages as the response and years of education and experience as predictors. Report and give a simple interpretation to the regression coefficient for years of education. Now fit the same model but with logged weekly wages. Give an interpretation to the regression coefficient for years of education. Which interpretation is more natural?

3. In this question, we investigate the relative merits of methods for computing the coefficients. Generate some artificial data by:

```
> x <- 1:20
> y <- x+rnorm(20)
```

   Fit a polynomial in x for predicting y. Compute $\hat{\beta}$ in two ways — by `lm()` and by using the direct calculation described in the chapter. At what degree of polynomial does the direct calculation method fail? (Note the need for the `I()` function in fitting the polynomial, that is, `lm(y ~ x + I(x^2))`.

4. The dataset `prostate` comes from a study on 97 men with prostate cancer who were due to receive a radical prostatectomy. Fit a model with `lpsa` as the response and `lcavol` as the predictor. Record the residual standard error and the R^2. Now add `lweight`, `svi`, `lpbh`, `age`, `lcp`, `pgg45` and `gleason` to the model one at a time. For each model record the residual standard error and the R^2. Plot the trends in these two statistics.

5. Using the `prostate` data, plot `lpsa` against `lcavol`. Fit the regressions of `lpsa` on `lcavol` and `lcavol` on `lpsa`. Display both regression lines on the plot. At what point do the two lines intersect?

CHAPTER 3

Inference

Until now, we have not found it necessary to assume any distributional form for the errors ε. However, if we want to make any confidence intervals or perform any hypothesis tests, we will need to do this. The common assumption is that the errors are normally distributed. In practice, this is often, although not always, a reasonable assumption. We have already assumed that the errors are independent and identically distributed (i.i.d.) with mean 0 and variance σ^2, so we have $\varepsilon \sim N(0, \sigma^2 I)$. Now since $y = X\beta + \varepsilon$, we have $y \sim N(X\beta, \sigma^2 I)$ which is a compact description of the regression model. From this we find, using the fact that linear combinations of normally distributed values are also normal, that:

$$\hat{\beta} = (X^T X)^{-1} X^T y \sim N(\beta, (X^T X)^{-1} \sigma^2)$$

3.1 Hypothesis Tests to Compare Models

Given several predictors for a response, we might wonder whether all are needed. Consider a larger model, Ω, and a smaller model, ω, which consists of a subset of the predictors that are in Ω. If there is not much difference in the fit, we would prefer the smaller model on the principle that simpler explanations are preferred. On the other hand, if the fit of the larger model is appreciably better, we will prefer it. We will take ω to represent the null hypothesis and Ω to represent the alternative. A geometrical view of the problem may be seen in Figure 3.1.

If $\text{RSS}_\omega - \text{RSS}_\Omega$ is small, then the fit of the smaller model is almost as good as the larger model and so we would prefer the smaller model on the grounds of simplicity. On the other hand, if the difference is large, then the superior fit of the larger model would be preferred. This suggests that something like:

$$\frac{\text{RSS}_\omega - \text{RSS}_\Omega}{\text{RSS}_\Omega}$$

would be a potentially good test statistic where the denominator is used for scaling purposes.

As it happens, the same test statistic arises from the likelihood-ratio testing approach. We give an outline of the development: If $L(\beta, \sigma | y)$ is the likelihood function, then the likelihood-ratio statistic is:

$$\frac{\max_{\beta, \sigma \in \Omega} L(\beta, \sigma | y)}{\max_{\beta, \sigma \in \omega} L(\beta, \sigma | y)}$$

The test should reject if this ratio is too large. Working through the details, we find

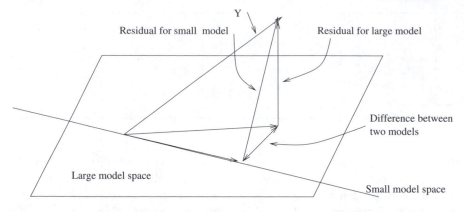

Figure 3.1 *Geometrical view of the comparison between big model, Ω, and small model, ω. The squared length of the residual vector for the big model is RSS_Ω while that for the small model is RSS_ω. By Pythagoras' theorem, the squared length of the vector connecting the two fits is $RSS_\omega - RSS_\Omega$. A small value for this indicates that the small model fits almost as well as the large model and thus might be preferred due to its simplicity.*

that for each model:

$$L(\hat{\beta}, \hat{\sigma}|y) \propto \hat{\sigma}^{-n}$$

which after some manipulation gives us a test that rejects if:

$$\frac{RSS_\omega - RSS_\Omega}{RSS_\Omega} > \text{a constant}$$

which is the same statistic suggested by the geometrical view. It remains for us to discover the null distribution of this statistic.

Now suppose that the dimension (or number of parameters) of Ω is p and the dimension of ω is q, then:

$$F = \frac{(RSS_\omega - RSS_\Omega)/(p-q)}{RSS_\Omega/(n-p)} \sim F_{p-q,n-p}$$

Details of the derivation of this statistic may be found in more theoretically oriented texts such as Sen and Srivastava (1990).

Thus we would reject the null hypothesis if $F > F_{p-q,n-p}^{(\alpha)}$. The degrees of freedom of a model are (usually) the number of observations minus the number of parameters so this test statistic can also be written:

$$F = \frac{(RSS_\omega - RSS_\Omega)/(df_\omega - df_\Omega)}{RSS_\Omega/df_\Omega}$$

where $df_\Omega = n - p$ and $df_\omega = n - q$. The same test statistic applies not just to when ω is a subset of Ω, but also to a subspace. This test is very widely used in regression and analysis of variance. When it is applied in different situations, the form of test statistic may be reexpressed in various different ways. The beauty of this approach is you only need to know the general form. In any particular case, you just need to figure out which models represent the null and alternative hypotheses, fit them and compute the test statistic. It is very versatile.

3.2 Testing Examples

Test of all the predictors

Are any of the predictors useful in predicting the response? Let the full model (Ω) be $y = X\beta + \varepsilon$ where X is a full-rank $n \times p$ matrix and the reduced model (ω) be $y = \mu + \varepsilon$. We would estimate μ by $\bar{y}$. We write the null hypothesis as:

$$H_0 : \beta_1 = \ldots \beta_{p-1} = 0$$

Now $\text{RSS}_\Omega = (y - X\hat{\beta})^T(y - X\hat{\beta}) = \hat{\varepsilon}^T\hat{\varepsilon}$, the residual sum of squares for the full model, while $\text{RSS}_\omega = (y - \bar{y})^T(y - \bar{y}) = TSS$, which is sometimes known as the sum of squares corrected for the mean. So in this case:

$$F = \frac{(\text{TSS} - \text{RSS})/(p-1)}{\text{RSS}/(n-p)}$$

We would now refer to $F_{p-1,n-p}$ for a critical value or a p-value. Large values of F would indicate rejection of the null. Traditionally, the information in the above test is presented in an *analysis of variance table*. Most computer packages produce a variant on this. See Table 3.1 for the layout. It is not really necessary to specifically compute all the elements of the table. As the originator of the table, Fisher said in 1931, it is "nothing but a convenient way of arranging the arithmetic." Since he had to do his calculations by hand, the table served a necessary purpose, but it is not essential now.

Source	Deg. of Freedom	Sum of Squares	Mean Square	F
Regression	$p-1$	SS_{reg}	$SS_{reg}/(p-1)$	F
Residual	$n-p$	RSS	$RSS/(n-p)$	
Total	$n-1$	TSS		

Table 3.1 *Analysis of variance table.*

A failure to reject the null hypothesis is not the end of the game — you must still investigate the possibility of nonlinear transformations of the variables and of outliers which may obscure the relationship. Even then, you may just have insufficient data to demonstrate a real effect, which is why we must be careful to say "fail to reject" the null rather than "accept" the null. It would be a mistake to conclude that no real relationship exists. This issue arises when a pharmaceutical company wishes to show that a proposed generic replacement for a brand-named drug is equivalent. It would not be enough in this instance just to fail to reject the null. A higher standard would be required.

When the null is rejected, this does not imply that the alternative model is the best model. We do not know whether all the predictors are required to predict the response or just some of them. Other predictors might also be added or existing predictors transformed or recombined. Either way, the overall F-test is just the beginning of an analysis and not the end.

Let's illustrate this test and others using an old economic dataset on 50 different countries. These data are averages from 1960 to 1970 (to remove business cycle or other short-term fluctuations). dpi is per capita disposable income in U.S. dollars; ddpi is the percentage rate of change in per capita disposable income; sr is aggregate personal saving divided by disposable income. The percentage of population under 15 (pop15) and over 75 (pop75) is also recorded. The data come from Belsley, Kuh, and Welsch (1980). Take a look at the data:

```
> data(savings)
> savings
                 sr pop15 pop75      dpi  ddpi
Australia     11.43 29.35  2.87 2329.68  2.87
Austria       12.07 23.32  4.41 1507.99  3.93
--- cases deleted ---
Malaysia       4.71 47.20  0.66  242.69  5.08
```

First, consider a model with all the predictors:

```
> g <- lm(sr ~ pop15 + pop75 + dpi + ddpi, savings)
> summary(g)
Coefficients:
             Estimate Std. Error t value Pr(>|t|)
(Intercept) 28.566087   7.354516    3.88  0.00033
pop15       -0.461193   0.144642   -3.19  0.00260
pop75       -1.691498   1.083599   -1.56  0.12553
dpi         -0.000337   0.000931   -0.36  0.71917
ddpi         0.409695   0.196197    2.09  0.04247

Residual standard error: 3.8 on 45 degrees of freedom
Multiple R-Squared: 0.338,      Adjusted R-squared: 0.28
F-statistic: 5.76 on 4 and 45 DF,  p-value: 0.00079
```

We can see directly the result of the test of whether any of the predictors have significance in the model. In other words, whether $\beta_1 = \beta_2 = \beta_3 = \beta_4 = 0$. Since the p-value, 0.00079, is so small, this null hypothesis is rejected.

We can also do it directly using the F-testing formula:

```
> g <- lm(sr ~ pop15 + pop75 + dpi + ddpi, savings)
> (tss <- sum((savings$sr-mean(savings$sr))^2))
[1] 983.63
> (rss <- deviance(g))
[1] 650.71
> df.residual(g)
[1] 45
> (fstat <- ((tss-rss)/4)/(rss/df.residual(g)))
[1] 5.7557
> 1-pf(fstat,4,df.residual(g))
[1] 0.00079038
```

Verify that the numbers match the regression summary above.

Testing just one predictor

Can one particular predictor be dropped from the model? The null hypothesis would be $H_0 : \beta_i = 0$. Let RSS_Ω be the RSS for the model with all the predictors of interest which has p parameters and let RSS_ω be the RSS for the model with all the same predictors except predictor i.

The F-statistic may be computed using the standard formula. An alternative approach is to use a t-statistic for testing the hypothesis:

$$t_i = \hat{\beta}_i / se(\hat{\beta}_i)$$

and check for significance using a t-distribution with $n - p$ degrees of freedom.

However, t_i^2 is exactly the F-statistic here, so the two approaches are numerically identical. The latter is less work and is presented in typical regression outputs.

For example, to test the null hypothesis that $\beta_1 = 0$, (that pop15 is not significant in the full model) we can simply observe that the p-value is 0.0026 from the table and conclude that the null should be rejected.

Let's do the same test using the general F-testing approach: We will need the RSS and df for the full model which are 650.71 and 45, respectively. We now fit the model that represents the null:

```
> g2 <- lm(sr ~ pop75 + dpi + ddpi, savings)
```

and compute the RSS and the F-statistic:

```
> (rss2 <- deviance(g2))
[1] 797.72
> (fstat <- (deviance(g2)-deviance(g))/
  (deviance(g)/df.residual(g)))
[1] 10.167
```

The p-value is then:

```
> 1-pf(fstat,1,df.residual(g))
[1] 0.002603
```

We can relate this to the t-based test and p-value by:

```
> sqrt(fstat)
[1] 3.1885
> (tstat <- summary(g)$coef[2,3])
[1] -3.1885
> 2*(1-pt(sqrt(fstat),45))
[1] 0.002603
```

A more convenient way to compare two nested models is:

```
> anova(g2,g)
Analysis of Variance Table

Model 1: sr ~ pop75 + dpi + ddpi
Model 2: sr ~ pop15 + pop75 + dpi + ddpi
  Res.Df Res.Sum Sq Df Sum Sq F value Pr(>F)
1     46        798
2     45        651  1    147    10.2 0.0026
```

Understand that this test of pop15 is relative to the other predictors in the model, namely, pop75, dpi and ddpi. If these other predictors were changed, the result of the test may be different. This means that it is not possible to look at the effect of pop15 in isolation. Simply stating the null hypothesis as $H_0 : \beta_{pop15} = 0$ is insufficient — information about what other predictors are included in the null is necessary. The result of the test may be different if the predictors change.

Testing a pair of predictors

Suppose we wish to test the significance of variables X_j and X_k. We might construct a table as seen above and find that both variables have p-values greater than 0.05 thus indicating that individually each one is not significant. Does this mean that both X_j and X_k can be eliminated from the model? *Not necessarily.*

Except in special circumstances, dropping one variable from a regression model causes the estimates of the other parameters to change so that we might find that after dropping X_j, a test of the significance of X_k shows that it should now be included in the model.

If you really want to check the joint significance of X_j and X_k, you should fit a model with and then without them and use the general F-test discussed before. Remember that even the result of this test may depend on what other predictors are in the model.

We test the hypothesis that both pop75 and ddpi may be excluded from the model:

```
> g3 <- lm(sr ~ pop15+dpi , savings)
> anova(g3,g)
Analysis of Variance Table

Model 1: sr ~ pop15 + dpi
Model 2: sr ~ pop15 + pop75 + dpi + ddpi
  Res.Df    RSS Df Sum of Sq      F  Pr(>F)
1     47 744.12
2     45 650.71  2     93.41 3.2299 0.04889
```

We see that the pair of predictors is just barely significant at the 5% level.

Tests of more than two predictors may be performed in a similar way by comparing the appropriate models.

Testing a subspace

Consider this example. Suppose that y is the first year grade point average for a student, X_j is the score on the quantitative part of a standardized test and X_k is the score on the verbal part. There might also be some other predictors. We might wonder whether we need two separate scores — perhaps they can be replaced by the total, $X_j + X_k$. So if the original model was:

$$y = \beta_0 + \cdots + \beta_j X_j + \beta_k X_k + \cdots + \varepsilon$$

then the reduced model is:

$$y = \beta_0 + \cdots + \beta_l (X_j + X_k) + \cdots + \varepsilon$$

which requires that $\beta_j = \beta_k$ for this reduction to be possible. So the null hypothesis is:

$$H_0 : \beta_j = \beta_k$$

This defines a linear subspace to which the general F-testing procedure applies. In our example, we might hypothesize that the effect of young and old people on the savings rate was the same or in other words that:

$$H_0 : \beta_{pop15} = \beta_{pop75}$$

In this case the null model would take the form:

$$y = \beta_0 + \beta_{dep}(pop15 + pop75) + \beta_{dpi}dpi + \beta_{ddpi}ddpi + \varepsilon$$

We can then compare this to the full model as follows:

```
> g <- lm(sr ~ .,savings)
> gr <- lm(sr ~ I(pop15+pop75)+dpi+ddpi,savings)
> anova(gr,g)
Analysis of Variance Table

Model 1: sr ~ I(pop15 + pop75) + dpi + ddpi
Model 2: sr ~ pop15 + pop75 + dpi + ddpi
  Res.Df Res.Sum Sq Df Sum Sq F value Pr(>F)
1     46        674
2     45        651  1     23    1.58   0.21
```

The period in the first model formula is shorthand for all the other variables in the data frame. The function `I()` ensures that the argument is evaluated rather than interpreted as part of the model formula. The p-value of 0.21 indicates that the null cannot be rejected here, meaning that there is not evidence that young and old people need to be treated separately in the context of this particular model.

Suppose we want to test whether one of the coefficients can be set to a particular value. For example:

$$H_0 : \beta_{ddpi} = 0.5$$

Here the null model would take the form:

$$y = \beta_0 + \beta_{pop15}pop15 + \beta_{pop75}pop75 + \beta_{dpi}dpi + 0.5ddpi + \varepsilon$$

Notice that there is now a fixed coefficient on the `ddpi` term. Such a fixed term in the regression equation is called an *offset*. We fit this model and compare it to the full:

```
> gr <- lm(sr ~ pop15+pop75+dpi+offset(0.5*ddpi),savings)
> anova(gr,g)
Analysis of Variance Table

Model 1: sr ~ pop15 + pop75 + dpi + offset(0.5 * ddpi)
Model 2: sr ~ pop15 + pop75 + dpi + ddpi
  Res.Df RSS Df Sum of Sq    F Pr(>F)
1     46 654
2     45 651  1         3 0.21   0.65
```

We see that the p-value is large and the null hypothesis here is not rejected. A simpler way to test such point hypotheses is to use a *t*-statistic:

$$t = (\hat{\beta} - c)/\text{se}(\hat{\beta})$$

where c is the point hypothesis. So in our example the statistic and corresponding p-value is:

```
> (tstat <- (0.409695-0.5)/0.196197)
[1] -0.46028
> 2*pt(tstat,45)
[1] 0.64753
```

We can see the p-value is the same as before and if we square the *t*-statistic

```
> tstat^2
[1] 0.21186
```

we find we get the same F-value as above. This latter approach is preferred in practice since we do not need to fit two models but it is important to understand that it is equivalent to the result obtained using the general F-testing approach.

Can we test a hypothesis such as $H_0 : \beta_j\beta_k = 1$ using our general theory? No. This hypothesis is not linear in the parameters so we cannot use our general method. We would need to fit a nonlinear model and that lies beyond the scope of this book.

3.3 Permutation Tests

We can put a different interpretation on the hypothesis tests we are making. For the Galápagos dataset, we might suppose that if the number of species had no relation to the five geographic variables, then the observed response values would be randomly distributed between the islands without relation to the predictors. The F-statistic is a good measure of the association between the predictors and the response with larger values indicating stronger associations. We might then ask what the chance would be under this assumption that an F-statistic would be observed as large, or larger than the one we actually observed. We could compute this exactly by computing the F-statistic for all possible ($n!$) permutations of the response variable and see what proportion exceed the observed F-statistic. This is a permutation test. If the observed proportion is small, then we must reject the contention that the response is unrelated to the predictors. Curiously, this proportion is estimated by the p-value calculated in the usual way based on the assumption of normal errors thus saving us from the massive task of actually computing the regression on all those computations. See Freedman and Lane (1983) for a discussion of these matters.

Let's see how we can apply the permutation test to the savings data. I chose a model with just pop75 and dpi so as to get a p-value for the F-statistic that is not too small (and therefore less interesting):

```
> g <- lm(sr ~ pop75+dpi,savings)
> summary(g)
Coefficients:
             Estimate Std. Error t value Pr(>|t|)
(Intercept) 7.056619   1.290435     5.47  1.7e-06
```

```
pop75          1.304965    0.777533    1.68      0.10
dpi           -0.000341    0.001013   -0.34      0.74
```

```
Residual standard error: 4.33 on 47 degrees of freedom
Multiple R-Squared: 0.102,        Adjusted R-squared: 0.0642
F-statistic: 2.68 on 2 and 47 DF,   p-value: 0.079
```

We can extract the F-statistic as:

```
> gs <- summary(g)
> gs$fstat
  value   numdf   dendf
 2.6796  2.0000 47.0000
```

The function sample() generates random permutations. We compute the F-statistic for 4000 randomly selected permutations and see what proportion exceeds the F-statistic for the original data:

```
> fstats <- numeric(4000)
> for(i in 1:4000){
+ ge <- lm(sample(sr) ~ pop75+dpi,data=savings)
+ fstats[i] <- summary(ge)$fstat[1]
+ }
> length(fstats[fstats > 2.6796])/4000
[1] 0.07425
```

This should take less than a minute on any relatively new computer. If you repeat this, you will get a slightly different result each time because of the random selection of the permutations. So our estimated p-value using the permutation test is 0.07425, which is close to the normal theory-based value of 0.0791. We could reduce variability in the estimation of the p-value simply by computing more random permutations. Since the permutation test does not depend on the assumption of normality, we might regard it as superior to the normal theory based value. It does take longer to compute, so we might use the normal inference secure, in the knowledge that the results can also be justified with an alternative argument.

Tests involving just one predictor also fall within the permutation test framework. We permute that predictor rather than the response. Let's test the pop75 predictor in the model. We can extract the t-statistic as:

```
> summary(g)$coef[2,3]
[1] 1.6783
```

Now we perform 4000 permutations of pop75 and check what fraction of the t-statistics exceeds 1.6783 in absolute value:

```
> tstats <- numeric(4000)
> for(i in 1:4000){
+ ge <- lm(sr ~ sample(pop75) + dpi, savings)
+ tstats[i] <- summary(ge)$coef[2,3]
+ }
> mean(abs(tstats) > 1.6783)
[1] 0.10475
```

The outcome is very similar to the observed normal-based p-value of 0.10.

3.4 Confidence Intervals for β

Confidence intervals (CIs) provide an alternative way of expressing the uncertainty in our estimates. They are linked to the tests that we have already constructed. For the CIs and regions that we will consider here, the following relationship holds. For a $100(1-\alpha)\%$ confidence region, any point that lies within the region represents a null hypothesis that would not be rejected at the $100\alpha\%$ level while every point outside represents a null hypothesis that would be rejected. So, in one sense, the confidence region provides a lot more information than a single hypothesis test in that it tells us the outcome of a whole range of hypotheses about the parameter values. Of course, by selecting the particular level of confidence for the region, we can only make tests at that level and we cannot determine the p-value for any given test simply from the region. However, since it is dangerous to read too much into the relative size of p-values (as far as how much evidence they provide against the null), this loss is not particularly important.

The confidence region tells us about plausible values for the parameters in a way that the hypothesis test cannot. This makes it more valuable. As with testing, we must decide whether to form confidence regions for parameters individually or simultaneously. Simultaneous regions are preferable, but for more than two dimensions they are difficult to display and so there is still some value in computing the one-dimensional CIs.

We can consider each parameter individually, which leads to CIs taking the general form of:

$$\text{Estimate} \pm \text{Critical Value} \times \text{SE of Estimate}$$

or specifically in this case:

$$\hat{\beta}_i \pm t_{n-p}^{(\alpha/2)} \hat{\sigma} \sqrt{(X^T X)_{ii}^{-1}}$$

Alternatively, a $100(1-\alpha)\%$ confidence region for β satisfies:

$$(\hat{\beta} - \beta)^T X^T X (\hat{\beta} - \beta) \leq p\hat{\sigma}^2 F_{p,n-p}^{(\alpha)}$$

These regions are ellipsoidally shaped. Because these ellipsoids live in higher dimensions, they cannot easily be visualized except for the two-dimensional case.

It is better to consider the joint CIs when possible, especially when the $\hat{\beta}$ are heavily correlated.

Consider the full model for the savings data:

```
> g <- lm(sr ~ ., savings)
```

where the output is displayed on page 28. We can construct individual 95% CIs for β_{pop75}:

```
> qt(0.975,45)
[1] 2.0141
> c(-1.69-2.01*1.08,-1.69+2.01*1.08)
[1] -3.8608  0.4808
```

CIs have a duality with two-sided hypothesis tests as we mentioned above. Thus the interval contains zero, which indicates that the null hypothesis $H_0 : \beta_{pop75} = 0$ would not be rejected at the 5% level. We can see from the summary that the p-value is 12.5% — greater than 5% — confirming this point.

The CI for β_{ddpi} is:

```
> c(0.41-2.01*0.196,0.41+2.01*0.196)
[1] 0.01604 0.80396
```

Because zero is not in this interval, the null is rejected. Nevertheless, this CI is relatively wide in the sense that the upper limit is about 50 times larger than the lower limit. This means that we are not really that confident about what the exact effect of growth on savings really is, even though it is statistically significant. A convenient way to obtain all the univariate intervals is:

```
> confint(g)
                  2.5 %       97.5 %
(Intercept) 13.7533307 43.3788424
pop15       -0.7525175 -0.1698688
pop75       -3.8739780  0.4909826
dpi         -0.0022122  0.0015384
ddpi         0.0145336  0.8048562
```

Now we construct the joint 95% confidence region for β_{pop15} and β_{pop75}. First, we load in a package for drawing confidence ellipses (which is not part of base R and so may need to be downloaded):

```
> library(ellipse)
```

and now the plot:

```
> plot(ellipse(g,c(2,3)),type="l",xlim=c(-1,0))
```

We add the origin and the point of the estimates:

```
> points(0,0)
> points(coef(g)[2],coef(g)[3],pch=18)
```

Since the origin lies outside the ellipse, we reject the hypothesis that $\beta_{pop15} = \beta_{pop75} = 0$. We mark the one-way CI on the plot for reference:

```
> abline(v=confint(g)[2,],lty=2)
> abline(h=confint(g)[3,],lty=2)
```

See the plot in Figure 3.2. Notice that these lines are not tangential to the ellipse. If the lines were moved out so that they enclosed the ellipse exactly, the CIs would be *jointly* correct.

In some circumstances, the origin could lie within both one-way CIs, but lie outside the ellipse. In this case, both one-at-a-time tests would not reject the null whereas the joint test would. The latter test would be preferred. It is also possible for the origin to lie outside the rectangle but inside the ellipse. In this case, the joint test would not reject the null whereas both one-at-a-time tests would reject. Again we prefer the joint test result.

Examine the correlation of the two predictors:

```
> cor(savings$pop15,savings$pop75)
[1] -0.90848
```

However, from the plot, we see that coefficients have a positive correlation. We can confirm this from:

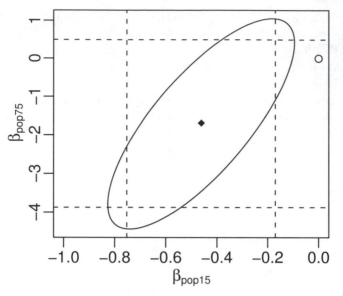

Figure 3.2 *Confidence ellipse and regions for* β_{pop75} *and* β_{pop15}

```
> summary(g,corr=TRUE)$corr
            (Intercept)     pop15      pop75       dpi       ddpi
(Intercept)     1.00000 -0.98416  -0.809111 -0.16588  -0.188265
pop15          -0.98416  1.00000   0.765356  0.17991   0.102466
pop75          -0.80911  0.76536   1.000000 -0.36705  -0.054722
dpi            -0.16588  0.17991  -0.367046  1.00000   0.255484
ddpi           -0.18827  0.10247  -0.054722  0.25548   1.000000
```

where we see that the correlation is 0.765. The correlation between predictors and the correlation between the coefficients of those predictors are often different in sign. Loosely speaking, two positively correlated predictors will attempt to perform the same job of explanation. The more work one does, the less the other needs to do and hence a negative correlation in the coefficients. For the negatively correlated predictors, as seen here, the effect is reversed.

3.5 Confidence Intervals for Predictions

Given a new set of predictors, x_0, the predicted response is $\hat{y}_0 = x_0^T \hat{\beta}$. However, we need to assess the uncertainty in this prediction. Decision makers need more than just a point estimate to make rational choices. If the prediction has a wide CI, we need to allow for outcomes far from the point estimate. For example, suppose we need to predict the high water mark of a river. We may need to construct barriers high enough to withstand floods much higher than the predicted maximum.

We must also distinguish between predictions of the future mean response and predictions of future observations. To make the distinction clear, suppose we have built a regression model that predicts the selling price of homes in a given area that is based on predictors such as the number of bedrooms and closeness to a major highway. There are two kinds of predictions that can be made for a given x_0:

1. Suppose a specific house comes on the market with characteristics x_0. Its selling price will be $x_0^T \beta + \varepsilon$. Since $E\varepsilon = 0$, the predicted price is $x_0^T \hat{\beta}$, but in assessing the variance of this prediction, we must include the variance of ε.

2. Suppose we ask the question — "What would a house with characteristics x_0 sell for on average?" This selling price is $x_0^T \beta$ and is again predicted by $x_0^T \hat{\beta}$ but now only the variance in $\hat{\beta}$ needs to be taken into account.

Most times, we will want the first case, which is called "prediction of a future value," while the second case, called "prediction of the mean response" is less commonly required.

We have var $(x_0^T \hat{\beta}) = x_0^T (X^T X)^{-1} x_0 \sigma^2$. A future observation is predicted to be $x_0^T \hat{\beta} + \varepsilon$ (where we do not know what the future ε will be and we can reasonably assume this to be independent of $\hat{\beta}$). So a $100(1-\alpha)$ % CI for a single future response is:

$$\hat{y}_0 \pm t_{n-p}^{(\alpha/2)} \hat{\sigma} \sqrt{1 + x_0^T (X^T X)^{-1} x_0}$$

If, on the other hand, you want a CI for the mean response for given x_0, use:

$$\hat{y}_0 \pm t_{n-p}^{(\alpha/2)} \hat{\sigma} \sqrt{x_0^T (X^T X)^{-1} x_0}$$

We return to the Galápagos data for this example:

```
> g <- lm(Species ~ Area+Elevation+Nearest+Scruz+Adjacent,gala)
```

Suppose we want to predict the number of species (of tortoise) on an island with predictors $0.08, 93, 6.0, 12.0, 0.34$ (same order as in the dataset). This might represent another island that was not included in the original data or we might wish to explore the effects of the predictors by experimenting with new predictor values. We can directly compute the point prediction as:

```
> x0 <- c(1,0.08,93,6.0,12.0,0.34)
> (y0 <- sum(x0*coef(g)))
[1] 33.92
```

This is the predicted number of species. If a whole number is preferred, we could round up to 34.

Now if we want a 95% CI for the prediction, we must decide whether we are predicting the number of species on one new island or the mean response for all islands with same predictors x_0. Suppose that an island was not surveyed for the original dataset. The former interval would be the one we want. For this dataset, the latter interval would be more valuable for "what if" type of calculations. First, we need the t-critical value:

```
> qt(0.975,24)
[1] 2.0639
```

We calculate the $(X^T X)^{-1}$ matrix:

```
> x <- model.matrix(g)
> xtxi <- solve(t(x) %*% x)
```

The width of the bands for mean response CI is ($\hat{\sigma} = 60.98$):

```
> (bm <- sqrt(x0 %*% xtxi %*% x0) *2.064 * 60.98)
      [,1]
[1,] 32.89
```

and the interval is:

```
> c(y0-bm,y0+bm)
[1]  1.0296 66.8097
```

Now we compute the prediction interval for the single future response:

```
> bm <- sqrt(1+x0 %*% xtxi %*% x0) *2.064 * 60.98
> c(y0-bm,y0+bm)
[1]  -96.17  164.01
```

Of course, the number of species cannot be negative. In such instances, impossible values in the CI can be avoided by transforming the response, say taking logs (explained in a later chapter), or by using a probability model more appropriate to the response. The normal distribution is supported on the whole real line and so negative values are always possible. A better choice for this example might be the Poisson distribution which is supported on the nonnegative integers.

There is a more direct method for computing the CI. The function predict() requires that its second argument be a data frame with variables named in the same way as the original dataset:

```
> x0 <- data.frame(Area=0.08,Elevation=93,Nearest=6.0,
  Scruz=12,Adjacent=0.34)
> str(predict(g,x0,se=TRUE))
List of 4
 $ fit            : num 33.9
 $ se.fit         : num 15.9
 $ df             : int 24
 $ residual.scale : num 61
> predict(g,x0,interval="confidence")
        fit    lwr    upr
[1,] 33.92 1.0338 66.806
> predict(g,x0,interval="prediction")
        fit     lwr    upr
[1,] 33.92 -96.153 163.99
```

Extrapolation occurs when we try to predict the response for values of the predictor which lie outside the range of the original data. There are two different types of extrapolation:

1. Quantitative extrapolation: We must check whether the new x_0 is within the range of the original data. If not, the prediction may be unrealistic. CIs for predictions get wider as we move away from the data. We can compute these bands for the Galápagos model where we vary the Nearest variable while holding the other predictors fixed:

```
> grid <- seq(0,100,1)
> p <- predict(g,data.frame(Area=0.08,Elevation=93,Nearest=
  grid,Scruz=12,Adjacent=0.34),se=T,interval="confidence")
> matplot(grid,p$fit,lty=c(1,2,2),type="l",
  xlab="Nearest",ylab="Species")
> rug(gala$Nearest)
```

We see that the confidence bands in Figure 3.3 become wider as we move away from the range of the data. However, this widening does not reflect the possibility that the structure of the model may change as we move into new territory. The uncertainty in the parametric estimates is allowed for, but not uncertainty about the model. The relationship may become nonlinear outside the range of the data — we have no way of knowing.

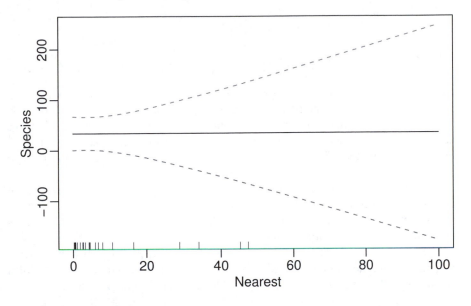

Figure 3.3 *Predicted* `Nearest` *over a range of values with 95% pointwise confidence bands for the mean response shown as dotted lines. A "rug" shows the location of the observed values of* `Nearest`.

2. Qualitative extrapolation: Is the new x_0 drawn from the same population from which the original sample was drawn? If the model was built in the past and is to be used for future predictions, we must make a difficult judgment as to whether conditions have remained constant enough for this to work.

3.6 Designed Experiments

In a designed experiment, the user has some control over X. For example, suppose we wish to compare several physical exercise regimes. The *experimental units* are the people we use for the study. We can choose some of the predictors such as the amount of time spent excercising or the type of equipment used. Some other predictors might not be controlled, but can be measured, such as the physical characteristics of the people. Other variables, such as the temperature in the room, might be held constant.

Our control over the conditions of the experiment allows us to make stronger conclusions from the analysis. Two important design features are *orthogonality* and *randomization*.

Orthogonality is a useful property because it allows us to more easily interpret the effect of one predictor without regard to another. Suppose we can partition X in two, $X = [X_1 | X_2]$ such that $X_1^T X_2 = 0$. So now:

$$Y = X\beta + \varepsilon = X_1\beta_1 + X_2\beta_2 + \varepsilon$$

and

$$X^T X = \begin{pmatrix} X_1^T X_1 & X_1^T X_2 \\ X_2^T X_1 & X_2^T X_2 \end{pmatrix} = \begin{pmatrix} X_1^T X_1 & 0 \\ 0 & X_2^T X_2 \end{pmatrix}$$

which means:

$$\hat{\beta}_1 = (X_1^T X_1)^{-1} X_1^T y \qquad \hat{\beta}_2 = (X_2^T X_2)^{-1} X_2^T y$$

Notice that $\hat{\beta}_1$ will be the same regardless of whether X_2 is in the model or not (and vice versa). So we can interpret the effect of X_1 without a concern for X_2. Unfortunately, the decoupling is not perfect for suppose we wish to test $H_0 : \beta_1 = 0$. We have RSS$/df = \hat{\sigma}^2$ that will be different depending on whether X_2 is included in the model or not, but the difference in F is not liable to be as large as in nonorthogonal cases.

Orthogonality is a desirable property, but will only occur when X is chosen by the experimenter. It is a feature of a good design. In observational data, we do not have direct control over X and this is the source of many of the interpretational difficulties associated with nonexperimental data.

Here is an example of an experiment to determine the effects of column temperature, gas/liquid ratio and packing height in reducing the unpleasant odor of a chemical product that was sold for household use. Read the data in and display:

```
> data(odor)
> odor
     odor temp gas pack
1     66   -1  -1    0
2     39    1  -1    0
3     43   -1   1    0
4     49    1   1    0
5     58   -1   0   -1
6     17    1   0   -1
7     -5   -1   0    1
8    -40    1   0    1
9     65    0  -1   -1
10     7    0   1   -1
11    43    0  -1    1
12   -22    0   1    1
13   -31    0   0    0
```

```
14   -35    0    0    0
15   -26    0    0    0
```

The three predictors have been transformed from their original scale of measurement, for example, temp = (Fahrenheit-80)/40 so the original values of the predictor were 40, 80 and 120. The data is presented in John (1971) and give an example of a *central composite* design. Here is the *X*-matrix:

```
> x <- as.matrix(cbind(1,odor[,-1]))
```

and $X^T X$:

```
> t(x) %*% x
        1 temp gas pack
   1 15    0   0    0
temp  0    8   0    0
 gas  0    0   8    0
pack  0    0   0    8
```

The matrix is diagonal. Even if temp was measured in the original Fahrenheit scale, the matrix would still be diagonal, but the entry in the matrix corresponding to temp would change. Now fit a model, while asking for the correlation of the coefficients:

```
> g <- lm(odor ~ temp + gas + pack, odor)
> summary(g,cor=T)
Coefficients:
              Estimate Std. Error t value Pr(>|t|)
(Intercept)       15.2        9.3    1.63     0.13
temp             -12.1       12.7   -0.95     0.36
gas              -17.0       12.7   -1.34     0.21
pack             -21.4       12.7   -1.68     0.12

Residual standard error: 36 on 11 degrees of freedom
Multiple R-Squared: 0.334,      Adjusted R-squared: 0.152
F-statistic: 1.84 on 3 and 11 DF,  p-value: 0.199

Correlation of Coefficients:
      (Intercept) temp gas
temp 0.00
gas  0.00         0.00
pack 0.00         0.00 0.00
```

We see that, as expected, the pairwise correlation of all the coefficients is zero. Notice that the SEs for the coefficients are equal due to the balanced design. Now drop one of the variables:

```
> g <- lm(odor ~ gas + pack, odor)
> summary(g)
Coefficients:
              Estimate Std. Error t value Pr(>|t|)
(Intercept)      15.20       9.26    1.64     0.13
gas             -17.00      12.68   -1.34     0.20
pack            -21.37      12.68   -1.69     0.12
```

```
Residual standard error: 35.9 on 12 degrees of freedom
Multiple R-Squared: 0.279,          Adjusted R-squared: 0.159
F-statistic: 2.32 on 2 and 12 DF,   p-value: 0.141
```

The coefficients themselves do not change, but the residual SE does change slightly, which causes small changes in the SEs of the coefficients, t-statistics and p-values, but nowhere near enough to change our qualitative conclusions.

These were data from an experiment so it was possible to control the values of the predictors to ensure orthogonality. Now consider the savings data, which are observational:

```
> g <- lm(sr ~ pop15 + pop75 + dpi + ddpi, savings)
> summary(g)
Coefficients:
              Estimate Std. Error t value Pr(>|t|)
(Intercept) 28.566087   7.354516    3.88  0.00033
pop15       -0.461193   0.144642   -3.19  0.00260
pop75       -1.691498   1.083599   -1.56  0.12553
dpi         -0.000337   0.000931   -0.36  0.71917
ddpi         0.409695   0.196197    2.09  0.04247

Residual standard error: 3.8 on 45 degrees of freedom
Multiple R-Squared: 0.338,          Adjusted R-squared: 0.28
F-statistic: 5.76 on 4 and 45 DF,   p-value: 0.00079
```

Drop pop15 from the model:

```
> g <- update(g, . ~ . - pop15)
> summary(g)


Coefficients:
              Estimate Std. Error t value Pr(>|t|)
(Intercept)  5.487494   1.427662    3.84  0.00037
pop75        0.952857   0.763746    1.25  0.21849
dpi          0.000197   0.001003    0.20  0.84499
ddpi         0.473795   0.213727    2.22  0.03162

Residual standard error: 4.16 on 46 degrees of freedom
Multiple R-Squared: 0.189,          Adjusted R-squared: 0.136
F-statistic: 3.57 on 3 and 46 DF,   p-value: 0.0209
```

Pay particular attention to pop75. The effect has now become positive whereas before it was negative. Granted, in both cases it is not significant, but it is not uncommon in other datasets for such sign changes to occur and for them to be significant.

Another important feature of designed experiments is the ability to randomly assign the experimental units to our chosen values of the predictors. For example, in our exercise regime experiment, we might choose an X with some desirable orthogonality properties and then fit a $y = X\beta + \varepsilon$ model. However, the subjects we use for the experiment are likely to differ in ways that might affect the response, which could, for example, be resting heart rate. Some of these characteristics may be known and measurable such as age or weight. These can be included in X. Including such known important variables is typically beneficial, as it increases the precision of our estimates of β.

However, there will be other variables, Z that we cannot measure or may not even suspect. The model we use is $y = X\beta + \varepsilon$ while the true model might be $y = X\beta + Z\gamma + \delta$. δ is the measurement error in the response. We can assume that $E\varepsilon = 0$ without any loss of generality, because if $E\varepsilon = c$, we could simply redefine β_0 as $\beta_0 + c$ and the error would again have expectation zero. This is another reason why it is generally unwise to remove the intercept term from the model since it acts as a sink for the mean effect of unincluded variables. So we see that ε incorporates both measurement error and the effect of other variables. We will not observe Z so we cannot estimate γ. If X were orthogonal to Z, that is, $X^T Z = 0$, the estimate of β would not depend on the presence of Z. Of course, we cannot achieve this exactly, but because of the random assignment of the experimental units, which each carry an unknown value of Z, to the predictors X, we will have $\text{cor}(X, Z) = 0$. This means that the estimate of β will not be biased by Z.

Another compelling argument in favor of the randomization is that it justifies the use of permutation tests. We really can argue that any assignment of X to the experimental units was one of $n!$ possible permutations.

For observational data, no randomization can be used in assigning X to the units and orthogonality will not just happen. An unmeasured and possible unsuspected *lurking* variable Z may be the real cause of an observed relationship between y and X. For example, we will observe a positive correlation among the shoe sizes and reading abilities of elementary school students, but this relationship is driven by a lurking variable — the age of the child. In this example, the nature of Z is obvious, but in other cases, we may not be able to identify an important Z. The possible existence of such a lurking variable casts a shadow over any conclusions from observational data.

Simply using a designed experiment does not ensure success. We may choose a bad model for the data or find that the variation in the unobserved Z overwhelms the effect of the predictors. Furthermore, results from unrealistic controlled experiments in the laboratory may not extrapolate to the real world.

3.7 Observational Data

Sometimes it is not practical or ethical to collect data from a designed experiment. We cannot control the assignment of X and so we can only obtain observational data. In some situations, we can control which cases we observe from those potentially available. A *sample survey* is used to collect the data. A good survey design can allow stronger and wider conclusions, but the data will still be observational.

Interpreting models built on observational data is problematic. There are many opportunities for error and any conclusions will carry substantial unquantifiable uncertainty. Nevertheless, there are many important questions for which only observational data will ever be available and so we must make the attempt in spite of the difficulties. Suppose we fit a model to obtain the regression equation:

$$\hat{y} = \hat{\beta}_0 + \hat{\beta}_1 x_1 + \cdots + \hat{\beta}_p x_p$$

What does $\hat{\beta}_1$ mean? In some cases, a β might represent a real physical constant. For example, we might attach weights x to a spring and measure the extension y. $\hat{\beta}_1$ will estimate a physical property of the spring. Such examples are rare and so usually the statistical model is just a convenience for representing a complex reality and the real meaning of a particular β is not obvious.

Let's start with the simplest interpretation: "A unit change in x_1 will produce a change of $\hat{\beta}_1$ in the response." For example, suppose y is annual income and x_1 is years of education. We might hope that $\hat{\beta}_1$ represents the predicted change in income if a particular individual had one more year of education.

The first objection is that there may be some lurking variable Z that is the real driving force behind y that also happens to be associated with x_1. Once Z is accounted for, there may be no relationship between x_1 and y. Unfortunately, we can usually never be certain that such a Z does not exist.

Even so, what if all relevant variables have been measured? In other words, suppose there are no unidentified lurking variables. Even then the naive interpretation does not work. Consider:

$$y = \hat{\beta}_0 + \hat{\beta}_1 x_1 + \hat{\beta}_2 x_2$$

but suppose we change $x_2 \rightarrow x_1 + x_2$; then:

$$y = \hat{\beta}_0 + (\hat{\beta}_1 - \hat{\beta}_2)x_1 + \hat{\beta}_2(x_1 + x_2)$$

The coefficient for x_1 has changed. Interpretation cannot be done separately for each variable. This is a practical problem because it is not unusual for the predictor of interest, x_1 in this example, to be mixed up in some way with other variables like x_2. This is the problem of *collinearity* which is explored in Section 5.3.

Let's try a new interpretation: "$\hat{\beta}_1$ is the effect of x_1 when all the other (specified) predictors are held constant."

This is better, but it too has problems. Often in practice, individual variables cannot be changed without changing others too. For example, in economics we cannot expect to change tax rates without other things changing too. Furthermore, this interpretation requires the specification of the other variables — changing which other variables are included will change the interpretation. Unfortunately, there is no simple solution.

Just to amplify this, consider the effect of pop75 on the savings rate in the savings dataset. I will fit four different models, all including pop75, but varying the inclusion of other variables:

```
> g <- lm(sr ~ pop15 + pop75 + dpi + ddpi, savings)
> summary(g)
Coefficients:
             Estimate Std. Error t value Pr(>|t|)
(Intercept) 28.566087   7.354516    3.88  0.00033
pop15       -0.461193   0.144642   -3.19  0.00260
pop75       -1.691498   1.083599   -1.56  0.12553
dpi         -0.000337   0.000931   -0.36  0.71917
ddpi         0.409695   0.196197    2.09  0.04247

Residual standard error: 3.8 on 45 degrees of freedom
```

```
Multiple R-Squared: 0.338,        Adjusted R-squared: 0.28
F-statistic: 5.76 on 4 and 45 DF,   p-value: 0.00079
```

It is perhaps surprising that pop75 is not significant in this model. However, pop75 is negatively correlated with pop15 since countries with proportionately more younger people are likely to have relatively fewer older ones and vice versa. These two variables both measure the nature of the age distribution in a country. When two variables that represent roughly the same thing are included in a regression equation, it is not unusual for one (or even both) of them to appear insignificant even though prior knowledge about the effects of these variables might lead one to expect them to be important:

```
> g2 <- lm(sr ~ pop75 + dpi + ddpi, savings)
> summary(g2)
Coefficients:
             Estimate Std. Error t value Pr(>|t|)
(Intercept) 5.487494   1.427662    3.84  0.00037
pop75       0.952857   0.763746    1.25  0.21849
dpi         0.000197   0.001003    0.20  0.84499
ddpi        0.473795   0.213727    2.22  0.03162

Residual standard error: 4.16 on 46 degrees of freedom
Multiple R-Squared: 0.189,        Adjusted R-squared: 0.136
F-statistic: 3.57 on 3 and 46 DF,   p-value: 0.0209
```

We note that the income variable dpi and pop75 are both not significant in this model and yet one might expect both of them to have something to do with savings rates. Higher values of these variables are both associated with wealthier countries. Let's see what happens when we drop dpi from the model:

```
> g3 <- lm(sr ~ pop75 + ddpi, savings)
> summary(g3)
Coefficients:
             Estimate Std. Error t value Pr(>|t|)
(Intercept)    5.470      1.410    3.88  0.00033
pop75          1.073      0.456    2.35  0.02299
ddpi           0.464      0.205    2.26  0.02856

Residual standard error: 4.12 on 47 degrees of freedom
Multiple R-Squared: 0.188,        Adjusted R-squared: 0.154
F-statistic: 5.45 on 2 and 47 DF,   p-value: 0.00742
```

Now pop75 is statistically significant with a positive coefficient. We try dropping ddpi:

```
> g4 <- lm(sr ~ pop75, savings)
> summary(g4)
Coefficients:
             Estimate Std. Error t value Pr(>|t|)
(Intercept)    7.152      1.248    5.73  6.4e-07
pop75          1.099      0.475    2.31  0.025

Residual standard error: 4.29 on 48 degrees of freedom
```

```
Multiple R-Squared:   0.1,          Adjusted R-squared: 0.0814
F-statistic: 5.34 on 1 and 48 DF,   p-value: 0.0251
```

The coefficient and p-value do not change much here due to the low correlation between `pop75` and `ddpi`. Compare the coefficients and p-values for `pop75` throughout. Notice how the sign and significance change in Table 3.2.

No. of Preds	Sign	Significant?
4	-	no
3	+	no
2	+	yes
1	+	yes

Table 3.2 *Sign and significance of* $\hat{\beta}_{pop75}$.

We see that the significance and the direction of the effect of `pop75` change according to what other variables are also included in the model. We see that no simple conclusion about the effect of `pop75` is possible. We must find interpretations for a variety of models. We certainly will not be able to make any strong causal conclusions.

In observational studies, there are steps one can take to make a stronger case for causality:

1. Try to include all relevant variables. If you have omitted an obvious important variable, critics of your study will find it easy to discount your results. A link between smoking and lung cancer was observed many years ago, but the evidence did not become overwhelming until many follow-up studies had discounted potential lurking variables. For example, smokers tend to drink more alcohol than nonsmokers. By including alcohol consumption in the model, we are able to adjust for its effect and observe any remaining effect due to smoking. Even so the possibility of a unsuspected lurking variable will always exist and we can never be absolutely sure.

2. Use nonstatistical knowledge of the physical nature of the relationship. For example, we can examine the lungs of smokers.

3. Try a variety of models and see whether a similar effect is observed. In complex data analyses involving several variables, we will usually find several models that fit the data well. Sometimes the variables included in these various models may be quite different, but if $\hat{\beta}_1$ is similar, this will increase confidence in the strength of the conclusions.

4. Multiple studies under different conditions can help confirm a relationship.

5. In a few cases, one can infer causality from an observational study. Dahl and Moretti (2003) report from U.S. census records that parents of a single girl are 5% more likely to divorce than parents of a single boy. The gap widens for larger families of girls and boys, respectively, and is even wider in several other countries. In many observational studies we can suggest a lurking variable that drives

both the predictor and the response, but here, excepting recent experimental repro-
ductive methods, the sex of a child is a purely random matter. This observational
study functions like an experimental design. So in this example, we can say that
the sex of the child affects the chance of divorce. The exact mechanism or reason
is unclear. Census data are so large that statistical significance is assured.

It is difficult to assess the evidence in these situations and one can never be certain.
The statistician is comfortable with the uncertainty expressed by hypothesis tests
and CIs, but the uncertainty associated with conclusions based on observational data
cannot be quantified and is necessarily subjective.

An alternative approach to interpreting parameter estimates is to recognize that
the parameters and their estimates are fictional quantities in most regression situations. The "true" values may never be known (if they even exist in the first place).
Instead, concentrate on predicting future values — these may actually be observed
and success can then be measured in terms of how good the predictions were.

Consider a prediction made using each of the previously mentioned four models:

```
> x0 <- data.frame(pop15=32,pop75=3,dpi=700,ddpi=3)
> predict(g,x0)
[1] 9.7267
> predict(g2,x0)
[1] 9.9055
> predict(g3,x0)
[1] 10.078
> predict(g4,x0)
[1] 10.448
```

Prediction is more stable than parameter estimation. This enables a rather cautious
interpretation of $\hat{\beta}_1$. Suppose the predicted value is $\hat{y}$ for given x_1 and other given
predictor values. Now suppose we *observe* $x_1 + 1$ and the same other given predictor
values; then the predicted response is increased by $\hat{\beta}_1$. Notice that we have been
careful not to say that we have taken a specific individual and increased his or her x_1
by 1; rather we have observed a new individual with predictor $x_1 + 1$. To put it another
way, people with college educations earn more on average than people without, but
giving a college education to someone without one will not necessarily increase his
or her income by the same amount.

3.8 Practical Difficulties

We have described a linear model $y = X\beta + \varepsilon$. Provided certain assumptions are satisfied, we can estimate β, test any linear hypothesis about β, construct confidence
regions for β and make predictions with CIs. The theory is clear and, with a little
experience, the computation becomes straightforward. However, the most difficult
part of regression data analysis is ensuring that the theory is appropriate for the real
application. Einstein put it well:

> So far as theories of mathematics are about reality; they are not certain; so far as they are
> certain, they are not about reality.

Problems may arise in several areas:

Nonrandom Samples

How the data were collected directly affects what conclusions we can draw. The general theory of hypothesis testing posits a *population* from which a *sample* is drawn — the sample is our data. We want to say something about the unknown *population* values β, using estimated values $\hat{\beta}$ that are obtained from the *sample* data. Furthermore, we require that the data be generated using a *simple random sample* of the population. This sample is finite in size, while the population is infinite in size or at least so large that the sample size is a negligible proportion of the whole. For more complex sampling designs, other procedures should be applied, but of greater concern is the case when the data are not a random sample at all.

Sometimes, researchers may try to select a *representative* sample by hand. Quite apart from the obvious difficulties in doing this, the logic behind the statistical inference depends on the sample being random. This is not to say that such studies are worthless, but that it would be unreasonable to apply anything more than descriptive statistical techniques. Confidence in the conclusions from such data is necessarily suspect.

A sample of convenience is where the data are not collected according to a sampling design. In some cases, it may be reasonable to proceed as if the data were collected using a random mechanism. For example, suppose we take the first 400 people from the phone book whose names begin with the letter P. Provided there is no ethnic effect, it may be reasonable to consider this a random sample from the population defined by the entries in the phone book. Here we are assuming the selection mechanism is effectively random with respect to the objectives of the study. The data are as good as random. Other situations are less clear-cut and judgment will be required. Such judgments are easy targets for criticism. Suppose you are studying the behavior of alcoholics and advertise in the media for study subjects. It seems very likely that such a sample will be biased, perhaps in unpredictable ways. In cases such as this, a sample of convenience is clearly biased in which case conclusions must be limited to the sample itself. This situation reduces to the case where the sample is the population.

Sometimes, the sample is the complete population. In this case, one might argue that inference is not required since the population and sample values are one and the same. For both regression datasets, `gala` and `savings`, that we have considered so far, the sample is effectively the population or a large and biased proportion thereof. Permutation tests make it possible to give some meaning to the p-value when the sample is the population or for samples of convenience although one has to be clear that one's conclusion applies only to the particular sample. Another approach that gives meaning to the p-value when the sample is the population involves the imaginative concept of "alternative worlds" where the sample/population at hand is supposed to have been randomly selected from parallel universes. This argument is definitely more tenuous.

Choice and Range of Predictors

When important predictors are not observed, the predictions may be poor or we may misinterpret the relationship between the predictors and the response.

The range and conditions under which the data are collected may limit effective predictions. It is unsafe to extrapolate too much. Carcinogen trials may apply large doses to mice. What do the results say about small doses applied to humans? Much of the evidence for harm from substances such as asbestos and radon comes from people exposed to much larger amounts than that encountered in a normal life. It is clear that workers in old asbestos manufacturing plants and uranium miners suffered from their respective exposures to these substances, but what does that say about the danger to you or me?

Model Misspecification

We make assumptions about the structural and random part of the model. For the error structure, we may assume that $\varepsilon \sim N(0, \sigma^2 I)$, but this may not be true. The structural part of linear model $Ey = X\beta$ may also be incorrect. The model we use may come from different sources:

1. Physical theory may suggest a model. For example, Hooke's law says that the extension of a spring is proportional to the weight attached. Models like these usually arise in the physical sciences and engineering.

2. Experience with past data. Similar data used in the past were modeled in a particular way. It is natural to see whether the same model will work with the current data. Models like these usually arise in the social sciences.

3. No prior idea exists — the model comes from an exploration of the data.

Confidence in the conclusions from a model declines as we progress through these. Models that derive directly from physical theory are relatively uncommon so that usually the linear model can only be regarded as an approximation to a complex reality.

The inference depends on the correctness of the model we use. We can partially check the assumptions about the model, but there will always be some element of doubt. Sometimes the data may suggest more than one possible model, which may lead to contradictory results.

Most statistical theory rests on the assumption that the model is correct. In practice, the best one can hope for is that the model is a fair representation of reality. A model can be no more than a good portrait. As George Box said,

> All models are wrong but some are useful.

Publication and Experimenter Bias

Many scientific journals will not publish the results of a study whose conclusions do not reject the null hypothesis. If different researchers keep studying the same relationship, sooner or later one of them will come up with a significant effect even if one really does not exist. It is not easy to find out about all the studies with negative results so it is easy to make the wrong conclusions. The news media often jump on the results of a single study, but one should be suspicious of these singleton results. Follow-up studies are often needed to confirm an effect.

Another source of bias is that researchers have a vested interest in obtaining a positive result. There is often more than one way to analyze the data and the researchers may be tempted to pick the one that gives them the results they want. This is not overtly dishonest, but it does lead to a bias towards positive results.

Practical and Statistical Significance

Statistical significance is not equivalent to practical significance. The larger the sample, the smaller your p-values will be, so do not confuse p-values with an important predictor effect. With large datasets it will be very easy to get statistically significant results, but the actual effects may be unimportant. Would we really care that test scores were 0.1% higher in one state than another or that some medication reduced pain by 2%? CIs on the parameter estimates are a better way of assessing the size of an effect. They are useful even when the null hypothesis is not rejected, because they tell us how confident we are that the true effect or value is close to the null.

It is also important to remember that a model is usually only an approximation of underlying reality which makes the exact meaning of the parameters debatable at the very least. The precision of the statement that $\beta_1 = 0$ exactly is at odds with the acknowledged approximate nature of the model. Furthermore, it is highly unlikely that a predictor that one has taken the trouble to measure and analyze has exactly zero effect on the response. It may be small but it will not be zero.

This means that in many cases, we know the point null hypothesis is false without even looking at the data. Furthermore, we know that the more data we have, the greater the power of our tests. Even small differences from zero will be detected with a large sample. Now if we fail to reject the null hypothesis, we might simply conclude that we did not have enough data to get a significant result. According to this view, the hypothesis test just becomes a test of sample size. For this reason, we prefer CIs.

Exercises

1. For the `prostate` data, fit a model with `lpsa` as the response and the other variables as predictors.

   (a) Compute 90 and 95% CIs for the parameter associated with `age`. Using just these intervals, what could we have deduced about the p-value for `age` in the regression summary?

   (b) Compute and display a 95% joint confidence region for the parameters associated with `age` and `lbph`. Plot the origin on this display. The location of the origin on the display tells us the outcome of a certain hypothesis test. State that test and its outcome.

   (c) Suppose a new patient with the following values arrives:

   ```
   lcavol   lweight        age       lbph         svi       lcp
   1.44692   3.62301  65.00000    0.30010    0.00000  -0.79851
   gleason      pgg45
   7.00000  15.00000
   ```

   Predict the `lpsa` for this patient along with an appropriate 95% CI.

(d) Repeat the last question for a patient with the same values except that he or she is age 20. Explain why the CI is wider.

(e) In the text, we made a permutation test corresponding to the F-test for the significance of all the predictors. Execute the permutation test corresponding to the t-test for age in this model. (Hint: {summary(g)$coef[4,3] gets you the t-statistic you need if the model is called g.)

2. For the model of the previous question, remove all the predictors that are not significant at the 5% level.

   (a) Recompute the predictions of the previous question. Are the CIs wider or narrower? Which predictions would you prefer? Explain.

   (b) Test this model against that of the previous question. Which model is preferred?

3. Using the teengamb data, fit a model with gamble as the response and the other variables as predictors.

   (a) Which variables are statistically significant?

   (b) What interpretation should be given to the coefficient for sex?

   (c) Predict the amount that a male with average (given these data) status, income and verbal score would gamble along with an appropriate 95% CI. Repeat the prediction for a male with maximal values (for this data) of status, income and verbal score. Which CI is wider and why is this result expected?

   (d) Fit a model with just income as a predictor and use an F-test to compare it to the full model.

4. Using the sat data:

   (a) Fit a model with total sat score as the response and expend, ratio and salary as predictors. Test the hypothesis that $\beta_{salary} = 0$. Test the hypothesis that $\beta_{salary} = \beta_{ratio} = \beta_{expend} = 0$. Do any of these predictors have an effect on the response?

   (b) Now add takers to the model. Test the hypothesis that $\beta_{takers} = 0$. Compare this model to the previous one using an F-test. Demonstrate that the F-test and t-test here are equivalent.

5. Find a formula relating R^2 and the F-test for the regression.

CHAPTER 4

Diagnostics

The estimation and inference from the regression model depends on several assumptions. These assumptions need to be checked using *regression diagnostics*. We divide the potential problems into three categories:

Error We have assumed that $\varepsilon \sim N(0, \sigma^2 I)$ or in words, that the errors are independent, have equal variance and are normally distributed.

Model We have assumed that the structural part of the model $Ey = X\beta$ is correct.

Unusual observations Sometimes just a few observations do not fit the model. These few observations might change the choice and fit of the model.

Diagnostic techniques can be graphical, which are more flexible but harder to definitively interpret, or numerical, which are narrower in scope, but require no intuition. The relative strengths of these two types of diagnostics will be explored below. The first model we try may prove to be inadequate. Regression diagnostics often suggest improvements, which means model building is often an iterative and interactive process. It is quite common to repeat the diagnostics on a succession of models.

4.1 Checking Error Assumptions

We wish to check the independence, constant variance and normality of the errors, ε. The errors are not observable, but we can examine the residuals, $\hat{\varepsilon}$. These are not interchangeable with the error, as they have somewhat different properties. Recall that $\hat{y} = X(X^T X)^{-1} X^T y = Hy$ where H is the hat-matrix, so that $\hat{\varepsilon} = y - \hat{y} = (I - H)y = (I - H)X\beta + (I - H)\varepsilon = (I - H)\varepsilon$. Therefore, var $\hat{\varepsilon} = $ var $(I - H)\varepsilon = (I - H)\sigma^2$ assuming that var $\varepsilon = \sigma^2 I$. We see that although the errors may have equal variance and be uncorrelated, the residuals do not. Fortunately, the impact of this is usually small and diagnostics are often applied to the residuals in order to check the assumptions on the error.

4.1.1 Constant Variance

It is not possible to check the assumption of constant variance just by examining the residuals alone — some will be large and some will be small, but this proves nothing. We need to check whether the variance in the residuals is related to some other quantity.

First, plot $\hat{\varepsilon}$ against $\hat{y}$. If all is well, you should see constant variance in the vertical ($\hat{\varepsilon}$) direction and the scatter should be symmetric vertically about zero. Things

to look for are heteroscedasticity (nonconstant variance) and nonlinearity (which indicates some change in the model is necessary). In Figure 4.1, these three cases are
illustrated.

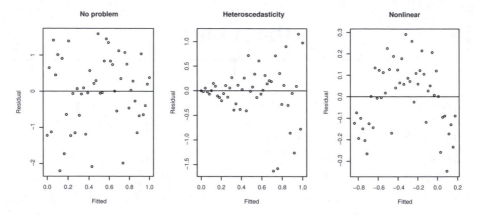

Figure 4.1 *Residuals vs. fitted plots — the first suggests no change to the current model
while the second shows nonconstant variance and the third indicates some nonlinearity, which
should prompt some change in the structural form of the model.*

You should also plot $\hat{\epsilon}$ against x_i (for predictors that are both in and out of the
model). Look for the same things except in the case of plots against predictors not in
the model, look for any relationship that might indicate that this predictor should be
included.

We illustrate this using the savings dataset:

```
> data(savings)
> g <- lm(sr ~ pop15+pop75+dpi+ddpi,savings)
```

First, the residuals vs. fitted plot and the absolute values of the residuals vs. fitted
plot:

```
> plot(fitted(g),residuals(g),xlab="Fitted",ylab="Residuals")
> abline(h=0)
> plot(fitted(g),abs(residuals(g)),
  xlab="Fitted",ylab="|Residuals|")
```

The plots may be seen in Figure 4.2. The latter plot is designed to check for nonconstant variance only. It folds over the bottom half of the first plot to increase the
resolution for detecting nonconstant variance. The first plot is still needed because
nonlinearity must be checked. We see no evidence of nonconstant variance.

A quick way to check nonconstant variance is this regression:

```
> summary(lm(abs(residuals(g)) ~ fitted(g)))
Coefficients:
             Estimate Std. Error t value Pr(>|t|)
(Intercept)     4.840      1.186    4.08  0.00017
fitted(g)      -0.203      0.119   -1.72  0.09250
```

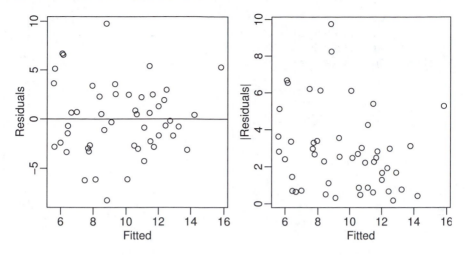

Figure 4.2 *Residual vs. fitted plots for the savings data.*

```
Residual standard error: 2.16 on 48 degrees of freedom
Multiple R-Squared: 0.0578,      Adjusted R-squared: 0.0382
F-statistic: 2.95 on 1 and 48 DF,  p-value: 0.0925
```

This test is not quite right, as some weighting should be used and the degrees of freedom should be adjusted, but there does not seem to be a clear problem with nonconstant variance.

It is often hard to judge residual plots without prior experience so it is helpful to generate some artificial plots where the true relationship is known. The following four for() loops show:

1. Constant variance

2. Strong nonconstant variance

3. Mild nonconstant variance

4. Nonlinearity

```
> par(mfrow=c(3,3))
> for(i in 1:9) plot(1:50,rnorm(50))
> for(i in 1:9) plot(1:50,(1:50)*rnorm(50))
> for(i in 1:9) plot(1:50,sqrt((1:50))*rnorm(50))
> for(i in 1:9) plot(1:50,cos((1:50)*pi/25)+rnorm(50))
> par(mfrow=c(1,1))
```

Repeat to get an idea of the usual amount of variation. Artificial generation of plots is a good way to "calibrate" diagnostic plots. It is often hard to judge whether an apparent feature is real or just random variation. Repeated generation of plots under a model, where there is no violation of the assumption that the diagnostic plot is designed to check, is helpful in making this judgment.

Now look at some residuals against predictor plots:

```
> plot(savings$pop15,residuals(g),
  xlab="Population under 15",ylab="Residuals")
> plot(savings$pop75,residuals(g),
  xlab="Population over 75",ylab="Residuals")
```

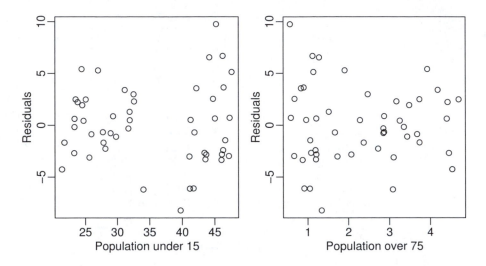

Figure 4.3 *Residuals vs. predictor plots for the savings data.*

The plots may be seen in Figure 4.3. Two groups can be seen in the first plot. Let's compare and test the variances in these groups. Given two independent samples from normal distributions, we can test for equal variance using the test statistic of the ratio of the two variance. The null distribution is an F with degrees of freedom given by the two samples:

```
> var.test(residuals(g)[savings$pop15>35],
  residuals(g)[savings$pop15<35])
```

```
        F test to compare two variances

F = 2.7851, num df = 22, denom df = 26, p-value = 0.01358
alternative hypothesis: true ratio of variances is not equal to 1
95 percent confidence interval:
 1.2410 6.4302
sample estimates:
ratio of variances
         2.7851
```

A significant difference is seen.

There are two approaches to dealing with nonconstant variance. Use of weighted least squares (see Section 6.1) is appropriate when the form of the nonconstant variance is either known exactly or there is some known parametric form. Alternatively, one can transform the variables. Sometimes other changes to the model may fix the problem, but we consider transforming the response y to $h(y)$ where $h()$ is chosen so that var $h(y)$ is constant. To see how to choose h consider this:

$$h(y) \;=\; h(Ey)+(y-Ey)h'(Ey)+\cdots$$
$$\text{var } h(y) \;=\; h'(Ey)^2 \text{var } y+\cdots$$

We ignore the higher order terms. For var $h(y)$ to be constant we need:

$$h'(Ey) \propto (\text{var } y)^{-1/2}$$

which suggests:

$$h(y)=\int \frac{dy}{\sqrt{\text{var } y}} == \int \frac{dy}{\text{SD} y}$$

For example if var $y =$ var $\varepsilon \propto (Ey)^2$, then $h(y) = \log y$ is suggested while if var $\varepsilon \propto (Ey)$, then $h(y) = \sqrt{y}$.

In practice, you need to look at the plot of the residuals and fitted values and take a guess at the relationship. When looking at the plot, we see the change in SDy rather then var y, because the SD is in the units of the response. If your initial guess is wrong, you can always try another transformation.

Sometimes it can be difficult to find a good transformation. For example, when $y_i \leq 0$ for some i, square root or log transformations will fail. You can try, say, $\log(y + \delta)$, for some small δ but this makes interpretation difficult.

Consider the residual vs. fitted plot for the Galápagos data:

```
> data(gala)
> gg <- lm(Species ~ Area + Elevation + Scruz + Nearest
+ Adjacent, gala)
> plot(fitted(gg),residuals(gg),xlab="Fitted",ylab="Residuals")
```

We can see nonconstant variance in the first plot of Figure 4.4.

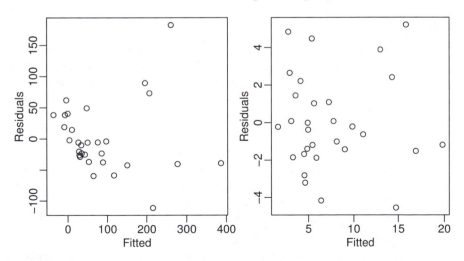

Figure 4.4 *Residual vs. fitted plots for the Galápagos data before (left) and after (right) transformation.*

There are formal tests for nonconstant variance — for example, one could start by regressing $|\hat{\varepsilon}|$ on $\hat{y}$ or x_i. The null hypothesis of constant variance is clear, but specifying a good alternative hypothesis is problematic. A formal test may be good at detecting a particular kind of nonconstant variance, but have no power to detect another. Residual plots are more versatile because unanticipated problems may be spotted.

A formal diagnostic test may have a reassuring aura of exactitude about it, but one needs to understand that any such test may be powerless to detect problems of an unsuspected nature. Graphical techniques are usually more effective at revealing structure that you may not have suspected. Of course, sometimes the interpretation of the plot may be ambiguous, but at least one can be sure that nothing is seriously wrong with the assumptions. For this reason, we usually prefer a graphical approach to diagnostics with formal tests reserved for the clarification of signs discovered in the plots.

Here we guess that a square root transformation will give us a constant variance:

```
> gs <- lm(sqrt(Species) ~ Area+ Elevation+ Scruz+ Nearest
  + Adjacent, gala)
> plot(fitted(gs),residuals(gs),xlab="Fitted",ylab="Residuals",
               main="Square root Response")
```

We see in the second plot of Figure 4.4 that the variance is now constant. Our guess at a variance stabilizing transformation worked out here, but had it not, we could always have tried something else. The square root transformation is often appropriate for count response data. The Poisson distribution is a good model for counts and that distribution has the property that the mean is equal to the variance thus suggesting the square root transformation.

4.1.2 Normality

The tests and confidence intervals we use are based on the assumption of normal errors. The residuals can be assessed for normality using a *Q–Q plot*. This compares the residuals to "ideal" normal observations. We plot the sorted residuals against $\Phi^{-1}\left(\frac{i}{n+1}\right)$ for $i = 1,\dots,n$.

Let's try it out on the savings data:

```
> qqnorm(residuals(g),ylab="Residuals")
> qqline(residuals(g))
```

See the first plot of Figure 4.5 — qqline() adds a line joining the first and third quartiles. It is not influenced by outliers. Normal residuals should follow the line approximately. Here, the residuals look normal.

Histograms and boxplots are not suitable for checking normality:

```
> hist(residuals(g))
```

The histogram seen in the second plot of Figure 4.5 does not have the expected bell shape. This is because we must group the data into bins. The choice of width and placement of these bins is problematic and the plot here is inconclusive.

We can get an idea of the variation to be expected in Q–Q plots in the following

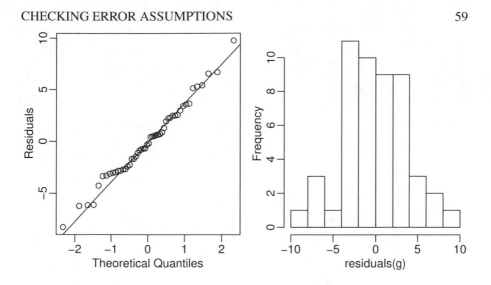

Figure 4.5 *Normality checks for the savings data.*

experiment. Such simulations are useful for self-calibration. I generate data from different distributions:

1. Normal

2. Lognormal — an example of a skewed distribution

3. Cauchy — an example of a long-tailed (leptokurtic) distribution

4. Uniform — an example of a short-tailed (platykurtic) distribution

Here is how to generate nine replicates at a time from each of these test cases:

```
> par(mfrow=c(3,3))
> for(i in 1:9) qqnorm(rnorm(50))
> for(i in 1:9) qqnorm(exp(rnorm(50)))
> for(i in 1:9) qqnorm(rcauchy(50))
> for(i in 1:9) qqnorm(runif(50))
> par(mfrow=c(1,1))
```

In Figure 4.6, you can see examples of all four cases:

It is not always easy to diagnose the problem in Q–Q plots. Sometimes extreme cases may be a sign of a long-tailed error like the Cauchy distribution or they can be just outliers. If removing such observations just results in other points becoming more prominent in the plot, the problem is likely due to a long-tailed error.

When the errors are not normal, least squares estimates may not be optimal. They will still be best linear unbiased estimates, but other *robust* estimators may be more effective. Also tests and confidence intervals are not exact. However, only long-tailed distributions cause large inaccuracies. Mild nonnormality can safely be ignored and the larger the sample size the less troublesome the nonnormality.

When nonnormality is found, the resolution depends on the type of problem found. For short-tailed distributions, the consequences of nonnormality are not serious and

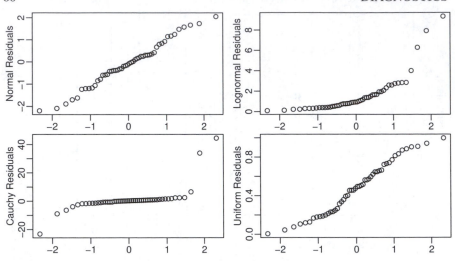

Figure 4.6 *Q–Q plots of simulated data.*

can reasonably be ignored. For skewed errors, a transformation of the response may solve the problem. For long-tailed errors, we might just accept the nonnormality and base the inference on the assumption of another distribution or use resampling methods such as the bootstrap or permutation tests. You do not want to do this unless absolutely necessary. Alternatively, use robust methods, which give less weight to outlying observations.

Also you may find that other diagnostics suggest changes to the model. In this changed model, the problem of nonnormal errors might not occur.

The Shapiro–Wilk test is a formal test for normality:

```
> shapiro.test(residuals(g))

        Shapiro-Wilk normality test

data:   residuals(g)
W = 0.987, p-value = 0.8524
```

The null hypothesis is that the the residuals are normal. Since the p-value is large, we do not reject this hypothesis.

We can only recommend this in conjunction with a Q–Q plot at best. The p-value is not very helpful as an indicator of what action to take. After all, with a large dataset, even mild deviations from nonnormality may be detected, but there would be little reason to abandon least squares because the effects of nonnormality are mitigated by large sample sizes. For smaller sample sizes, formal tests lack power.

4.1.3 Correlated Errors

We assume that the errors are uncorrelated, but for temporally or spatially related data this may well be untrue. For this type of data, it is wise to check the uncorrelated assumption.

Graphical checks include plots of $\hat{\varepsilon}$ against time and $\hat{\varepsilon}_i$ against $\hat{\varepsilon}_{i-1}$, while the Durbin–Watson test uses the statistic:

$$DW = \frac{\sum_{i=2}^{n}(\hat{\varepsilon}_i - \hat{\varepsilon}_{i-1})^2}{\sum_{i=1}^{n}\hat{\varepsilon}_i^2}$$

The null distribution based on the assumption of uncorrelated errors follows a linear combination of χ^2 distributions. The test is implemented in the lmtest package. The run test is an alternative.

For the example, we use some data taken from an environmental study that measured four variables — ozone, radiation, temperature and wind speed — for 153 consecutive days in New York:

```
> data(airquality)
> airquality
    Ozone Solar.R Wind Temp Month Day
1      41     190  7.4   67     5   1
2      36     118  8.0   72     5   2
3      12     149 12.6   74     5   3
4      18     313 11.5   62     5   4
5      NA      NA 14.3   56     5   5
etc..
```

We notice that there are some missing values. Take a look at the data (plot not shown):

```
> pairs(airquality,panel=panel.smooth)
```

We fit a standard linear model and check the residual vs. fitted plot in Figure 4.7.

```
> g <- lm(Ozone ~ Solar.R + Wind + Temp,airquality,
    na.action = na.exclude)
> summary(g)
Coefficients:
              Estimate Std. Error t value Pr(>|t|)
(Intercept) -64.3421    23.0547   -2.79   0.0062
Solar.R       0.0598     0.0232    2.58   0.0112
Wind         -3.3336     0.6544   -5.09  1.5e-06
Temp          1.6521     0.2535    6.52  2.4e-09

Residual standard error: 21.2 on 107 degrees of freedom
Multiple R-Squared: 0.606,      Adjusted R-squared: 0.595
F-statistic: 54.8 on 3 and 107 DF,   p-value: <2e-16
> plot(fitted(g),residuals(g),xlab="Fitted",ylab="Residuals")
```

Notice how there are only 107 degrees corresponding to the 111 complete observations. The default behavior in R when performing a regression with missing values is to omit any case that contains a missing value. The option na.action =

na.exclude does not use cases with missing values in the computation, but does keep track of which cases are missing in the residuals, fitted values and other quantities. We see some nonconstant variance and nonlinearity and so we try transforming the response:

```
> g1 <- lm(log(Ozone) ~ Solar.R + Wind + Temp,
  airquality,na.action=na.exclude)
> plot(fitted(g1),residuals(g1),xlab="Fitted",ylab="Residuals")
```

The improvement can be seen in the second panel of Figure 4.7.

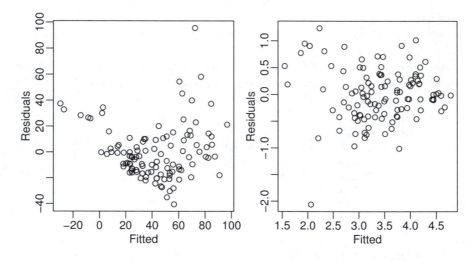

Figure 4.7 *Residuals vs. fitted for the air quality data. Untransformed response on the left; logged response on the right.*

Now we check the residuals for correlation. First, make an index plot of the residuals — see the first plot of Figure 4.8:

```
> plot(residuals(g1),ylab="Residuals")
> abline(h=0)
```

If there was serial correlation, we would see either longer runs of residuals above or below the line for positive correlation or greater than normal fluctuation for negative correlations. Unless these effects are strong, they can be difficult to spot. Nothing is obviously wrong here. It is often better to plot successive residuals:

```
> plot(residuals(g1)[-153],residuals(g1)[-1], xlab=
  expression(hat(epsilon)[i]),ylab=expression(hat(epsilon)[i+1]))
```

There is no obvious problem with correlation here. We can see a single outlier (which gets plotted twice in this case). Let's check using a regression of successive residuals — the intercept is omitted because residuals have mean zero:

```
> summary(lm(residuals(g1)[-1] ~ -1+residuals(g1)[-153]))
Coefficients:
```

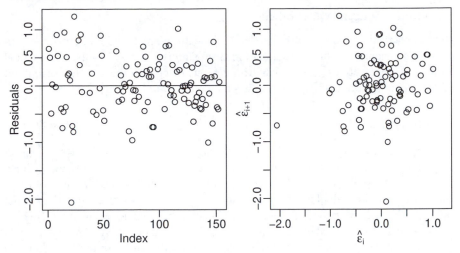

Figure 4.8 *Diagnostic plots for correlated errors in the air quality data.*

```
                     Estimate Std. Error t value Pr(>|t|)
residuals(gl)[-153]    0.110      0.105    1.05      0.3

Residual standard error: 0.508 on 91 degrees of freedom
Multiple R-Squared: 0.0119,     Adjusted R-squared: 0.00107
F-statistic:  1.1 on 1 and 91 DF,  p-value: 0.297
```

We omitted the intercept term because the residuals have mean zero. We see that there is no significant correlation. You can plot more than just successive pairs if you suspect a more complex dependence.

We can compute the Durbin–Watson statistic:

```
> library(lmtest)
> dwtest(Ozone ~ Solar.R + Wind + Temp,data=na.omit(airquality))

        Durbin-Watson test

data:  Ozone ~ Solar.R + Wind + Temp
DW = 1.9355, p-value = 0.3347
alternative hypothesis: true autocorrelation is greater than 0
```

where the p-value indicates no evidence of correlation. However, the result should be viewed with skepticism because of our omission of the missing values.

If you do have correlated errors, you can use generalized least squares — see Chapter 6. For data where there is no apparent link between observations, as there is in serial data, it is almost impossible to check for correlation between errors. Fortunately, there is no reason to suspect it either.

4.2 Finding Unusual Observations

Some observations do not fit the model well — these are called outliers. Other observations change the fit of the model in a substantive manner — these are called influential observations. A point can be none, one or both of these. A leverage point is unusual in the predictor space — it has the potential to influence the fit.

4.2.1 Leverage

$h_i = H_{ii}$ are called *leverages* and are useful diagnostics. Since var $\hat{\varepsilon}_i = \sigma^2(1 - h_i)$, a large leverage, h_i, will make var $\hat{\varepsilon}_i$ small. The fit will be "forced" close to y_i. Since $\sum_i h_i = p$, an average value for h_i is p/n. A "rule of thumb" is that leverages of more than $2p/n$ should be looked at more closely. Large values of h_i are due to extreme values in X. h_i corresponds to a (squared) Mahalanobis distance defined by X which is $(x - \bar{x})^T \hat{\Sigma}^{-1}(x - \bar{x})$ where $\hat{\Sigma}$ is the estimated covariance of X. The value of h_i depends only on X and not y so leverages contain only partial information about a point.

We will use the savings dataset as an example here:

```
> g <- lm(sr ~ pop15 + pop75 + dpi + ddpi, savings)
> ginf <- influence(g)
> ginf$hat
 [1] 0.067713 0.120384 0.087482 0.089471 0.069559 0.158402
....
> sum(ginf$hat)
[1] 5
```

We verify that the sum of the leverages is indeed five — the number of parameters in the model.

Without making assumptions about the distributions of the predictors that would often be unreasonable, we cannot say how the leverages would be distributed. Nevertheless, we would like to identify unusually large values of the leverage. The half-normal plot is a good way to do this.

Half-normal plots are designed for the assessment of positive data. They could be used for $|\hat{\varepsilon}|$, but are more typically useful for diagnostic quantities like the leverages. The idea is to plot the data against the positive normal quantiles.

The steps are:

1. Sort the data: $x_{[1]} \leq \ldots x_{[n]}$.

2. Compute $u_i = \Phi^{-1}\left(\frac{n+i}{2n+1}\right)$.

3. Plot $x_{[i]}$ against u_i.

We are usually not looking for a straight line relationship since we do not necessarily expect a positive normal distribution for quantities like the leverages. We are looking for outliers, which will be apparent as points that diverge substantially from the rest of the data.

We demonstrate the half-normal plot on the leverages for the savings data:

```
> countries <- row.names(savings)
> halfnorm(lm.influence(g)$hat,labs=countries,ylab="Leverages")
```

The plot is the first shown in Figure 4.9 — I have plotted the country name instead of just a dot for the largest two cases, respectively, to aid identification:

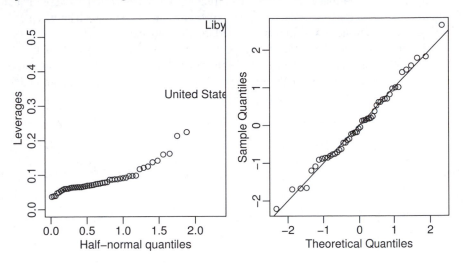

Figure 4.9 *Half-normal plots for the leverages and a Q–Q plot for the studentized residuals.*

As we have seen var $\hat{\varepsilon}_i = \sigma^2(1 - h_i)$ this suggests the use of:

$$r_i = \frac{\hat{\varepsilon}_i}{\hat{\sigma}\sqrt{1 - h_i}}$$

which are called (internally) *studentized residuals*. If the model assumptions are correct, var $r_i = 1$ and $corr(r_i, r_j)$ tends to be small. Studentized residuals are sometimes preferred in residual plots, as they have been standardized to have equal variance. Studentization can only correct for the natural nonconstant variance in residuals when the errors have constant variance. If there is some underlying heteroscedasticity in the errors, studentization cannot correct for it.

We now get the studentized residuals for the savings data:

```
> gs <- summary(g)
> gs$sig
[1] 3.8027
> stud <- residuals(g)/(gs$sig*sqrt(1-ginf$hat))
> qqnorm(stud)
> abline(0,1)
```

We have displayed the Q–Q plot of the studentized residuals in the second plot of Figure 4.9. Because these residuals have been standardized, we expect the points to approximately follow the $y = x$ line if normality holds.

Some authors recommend using studentized rather than raw residuals in all diagnostic plots. However, in many cases, the studentized residuals are not effectively very different from the raw residuals. Only when there is unusually large leverage will the differences be noticeable.

4.2.2 Outliers

An outlier is a point that does not fit the current model. We need to be aware of such exceptions. An outlier test is useful because it enables us to distinguish between truly unusual observations and residuals that are large, but not exceptional.

Outliers may affect the fit. See Figure 4.10. The two additional marked points both have high leverage because they are far from the rest of the data. ▲ is not an outlier. ● does not have a large residual if it is included in the fit. Only when we compute the fit without that point do we get a large residual.

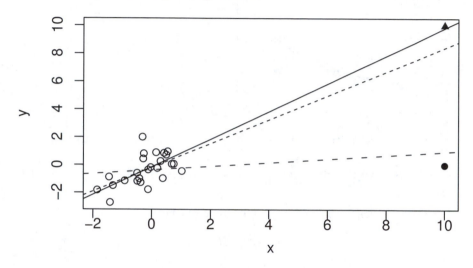

Figure 4.10 *Outliers can conceal themselves. The solid line is the fit including the* ▲ *point but not the* ● *point. The dotted line is the fit without either additional point and the dashed line is the fit with the* ● *point but not the* ▲ *point.*

To detect such points, we exclude point i and recompute the estimates to get $\hat{\beta}_{(i)}$ and $\hat{\sigma}^2_{(i)}$ where (i) denotes that the ith case has been excluded. Hence:

$$\hat{y}_{(i)} = x_i^T \hat{\beta}_{(i)}$$

If $\hat{y}_{(i)} - y_i$ is large, then case i is an outlier. Just looking at $\hat{\varepsilon}_i$ misses those difficult observations which pull the regression line so close to them that they conceal their true status. To judge the size of a potential outlier, we need an appropriate scaling. We find:

$$\text{vâr}\,(y_i - \hat{y}_{(i)}) = \hat{\sigma}^2_{(i)}(1 + x_i^T\,(X_{(i)}^T X_{(i)}))^{-1}x_i)$$

and so we define the jackknife (or externally studentized or crossvalidated) residuals as:

$$t_i = \frac{y_i - \hat{y}_{(i)}}{\hat{\sigma}_{(i)}(1 + x_i^T\,(X_{(i)}^T X_{(i)}))^{-1}x_i)^{1/2}}$$

which are distributed t_{n-p-1} if the model is correct and $\varepsilon \sim N(0, \sigma^2 I)$. Fortunately, there is an easier way to compute t_i:

$$t_i = \frac{\hat{\varepsilon}_i}{\hat{\sigma}_{(i)}\sqrt{1-h_i}} = r_i \left(\frac{n-p-1}{n-p-r_i^2}\right)^{1/2}$$

which avoids doing n regressions.

Since $t_i \sim t_{n-p-1}$, we can calculate a p-value to test whether case i is an outlier. Even though we might explicity test only one or two large t_is, by identifying them as large, we are implicitly testing all cases. Some adjustment of the level of the test is necessary; otherwise we would identify around 5% of observations as outliers even when none exist.

Suppose we want a level α test. Now P(all tests accept) = 1− P(at least one rejects) $\geq 1 - \sum_i P(\text{test } i \text{ rejects}) = 1 - n\alpha$. So this suggests that if an overall level α test is required, then a level α/n should be used in each of the tests. This method is called the *Bonferroni correction* and is used in contexts other than outliers also. Its biggest drawback is that it is conservative — it finds fewer outliers than the nominal level of confidence would dictate. The larger that n is, the more conservative it gets.

Now get the jackknife residuals for the savings data:

```
> jack <- rstudent(g)
> jack[which.max(abs(jack))]
Zambia
2.8536
```

The largest residual of 2.85 is pretty big for a standard normal scale, but is it an outlier? Compute the Bonferroni critical value:

```
> qt(.05/(50*2),44)
[1] -3.5258
```

Since 2.85 is less than 3.52, we conclude that Zambia is *not* an outlier. For simple regression, the minimum critical value occurs at $n = 23$ taking the value 3.51. This indicates that it is not worth the trouble of computing the outlier test p-value unless the jackknife residual exceeds about 3.5 in absolute value.

Some points to consider about outliers:

1. Two or more outliers next to each other can hide each other.

2. An outlier in one model may not be an outlier in another when the variables have been changed or transformed. You will usually need to reinvestigate the question of outliers when you change the model.

3. The error distribution may not be normal and so larger residuals may be expected. For example, day-to-day changes in stock indices seem mostly normal, but large changes occur frequently.

4. Individual outliers are usually much less of a problem in larger datasets. A single point will not have the leverage to affect the fit very much. It is still worth identifying outliers if these types of observations are worth knowing about in the particular application. For large datasets, we need only to worry about clusters of outliers. Such clusters are less likely to occur by chance and more likely to represent actual structure. Finding these clusters is not always easy.

What should be done about outliers?

1. Check for a data-entry error first. These are relatively common. Unfortunately, the original source of the data may have been lost.

2. Examine the physical context — why did it happen? Sometimes, the discovery of an outlier may be of singular interest. Some scientific discoveries spring from noticing unexpected aberrations. Another example of the importance of outliers is in the statistical analysis of credit card transactions. Outliers in this case may represent fraudulent use.

3. Exclude the point from the analysis but try reincluding it later if the model is changed. The exclusion of one or more observations may make the difference between getting a statistically significant result or having some unpublishable research. This can lead to a difficult decision about what exclusions are reasonable. To avoid any suggestion of dishonesty, always report the existence of outliers even if you do not include them in your final model.

4. Suppose you find outliers that cannot reasonably be identified as mistakes or aberrations, but are viewed as naturally occurring. Rather than exclude these points and then use least squares, it is more efficient and reliable to use robust regression, as explained in Section 6.4. The preference for robust regression becomes stronger when there are multiple outliers. Outlier rejection in conjunction with least squares is not a good method of estimation.

5. It is dangerous to exclude outliers in an automatic manner. National Aeronautics and Space Administation (NASA) launched the *Nimbus 7* satellite to record atmospheric information. After several years of operation in 1985, the British Antarctic Survey observed a large decrease in atmospheric ozone over the Antarctic. On further examination of the NASA data, it was found that the data processing program automatically discarded observations that were extremely low and assumed to be mistakes. Thus the discovery of the Antarctic ozone hole was delayed several years. Perhaps, if this had been known earlier, the chlorofluorocarbon (CFC) phaseout would have been agreed upon earlier and the damage could have been limited. See Stolarski et al. (1986) for more.

Here is an example of a dataset with multiple outliers. Data are available on the log of the surface temperature and the log of the light intensity of 47 stars in the star cluster CYG OB1, which is in the direction of Cygnus. These data appear in Rousseeuw and Leroy (1987).

Read in and plot the data:

```
> data(star)
> plot(star$temp,star$light,xlab="log(Temperature)",
  ylab="log(Light Intensity)")
```

There appears to be a positive correlation between temperature and light intensity, but there are four stars that do not fit the pattern. We fit a linear regression and add the fitted line to the plot:

```
> ga <- lm(light ~ temp, star)
> abline(ga)
```

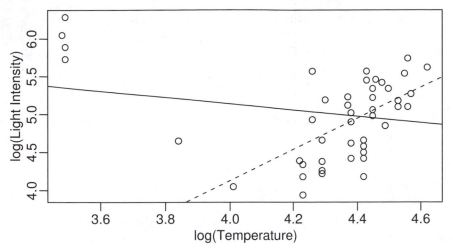

Figure 4.11 *Regression line including four leftmost points is solid and excluding these points is dotted.*

The plot is seen in Figure 4.11 with the regression line in solid type. This line does not follow the bulk of the data because it tries to fit the four unusual points. We check whether the outlier test detects these points:

```
> range(rstudent(ga))
[1] -2.0494  1.9058
```

No outliers are found even though we can see them clearly in the plot. The four stars on the upper left of the plot are giants. See what happens if these are excluded:

```
> ga <- lm(light ~ temp, data=star, subset=(temp>3.6))
> abline(ga,lty=2)
```

This illustrates the problem of multiple outliers. We can visualize the problems here and take corrective action, but for higher dimensional data this is much more difficult. Robust regression methods would be superior here.

4.2.3 Influential Observations

An influential point is one whose removal from the dataset would cause a large change in the fit. An influential point may or may not be an outlier and may or may not have large leverage but it will tend to have at least one of these two properties. In Figure 4.10, the ▲ point is not an influential point but the ● point is.

There are several measures of influence. A subscripted (i) indicates the fit without case i. We might consider the change in the fit $X^T(\hat{\beta} - \hat{\beta}_{(i)}) = \hat{y} - \hat{y}_{(i)}$, but there will be n of these length n vectors to examine. For a more compact diagnostic, we might consider the change in the coefficients $\hat{\beta} - \hat{\beta}_{(i)}$. There will be $n \times p$ of these to look at. The Cook statistics are a popular influence diagnostic because they reduce the information to a single value for each case. They are defined as:

$$D_i \;=\; \frac{(\hat{y} - \hat{y}_{(i)})^T (\hat{y} - \hat{y}_{(i)})}{p\hat{\sigma}^2}$$

$$=\; \frac{1}{p} r_i^2 \frac{h_i}{1 - h_i}$$

The first term, r_i^2, is the residual effect and the second is the leverage. The combination of the two leads to influence. A half-normal plot of D_i can be used to identify influential observations.

Continuing with our study of the savings data:

```
> cook <- cooks.distance(g)
> halfnorm(cook,3,labs=countries,ylab="Cook's distances")
```

The Cook statistics may be seen in the first plot of Figure 4.12. I have identified the largest three values. We now exclude the largest one (Libya) and see how the fit

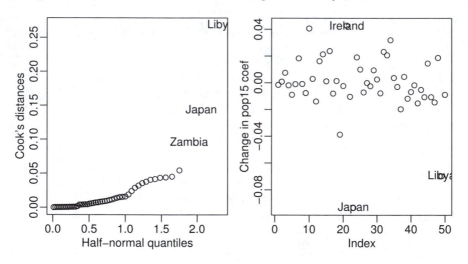

Figure 4.12 *Half-normal plot of the Cook statistics and* $\hat{\beta} - \hat{\beta}_{(i)}$*'s for* pop15 *for the savings data.*

changes:

```
> g1 <- lm(sr ~ pop15+pop75+dpi+ddpi,savings,
    subset=(cook < max(cook)))
> summary(g1)
Coefficients:
              Estimate  Std. Error  t value  Pr(>|t|)
(Intercept) 24.524046    8.224026     2.98    0.0047
pop15       -0.391440    0.157909    -2.48    0.0171
pop75       -1.280867    1.145182    -1.12    0.2694
dpi         -0.000319    0.000929    -0.34    0.7331
ddpi         0.610279    0.268778     2.27    0.0281
```

```
Residual standard error: 3.79 on 44 degrees of freedom
Multiple R-Squared: 0.355,        Adjusted R-squared: 0.297
F-statistic: 6.07 on 4 and 44 DF,  p-value: 0.000562
```

Compared to the full data fit:

```
> summary(g)
Coefficients:
              Estimate Std. Error t value Pr(>|t|)
(Intercept) 28.566087   7.354516    3.88  0.00033
pop15       -0.461193   0.144642   -3.19  0.00260
pop75       -1.691498   1.083599   -1.56  0.12553
dpi         -0.000337   0.000931   -0.36  0.71917
ddpi         0.409695   0.196197    2.09  0.04247
```

```
Residual standard error: 3.8 on 45 degrees of freedom
Multiple R-Squared: 0.338,        Adjusted R-squared: 0.28
F-statistic: 5.76 on 4 and 45 DF,  p-value: 0.00079
```

Among other changes, we see that the coefficient for ddpi changed by about 50%. We do not like our estimates to be so sensitive to the presence of just one country. It would be rather tedious to leave out each country in turn, so we examine the leave-out-one differences in the coefficients:

```
> plot(ginf$coef[,2],ylab="Change in pop15 coef")
> identify(1:50,ginf$coef[,2],countries)
```

We just plotted the change in the second parameter estimate, $(\hat{\beta}_{pop15})$ when a case is left out, as seen in the second panel of Figure 4.12. The identify function allows interactive identification of points by clicking the left mouse button on the plot and then using the middle mouse button to finish. This plot should be repeated for the other variables. Japan sticks out on this particular plot so examine the effect of removing it:

```
> gj <- lm(sr ~ pop15+pop75+dpi+ddpi,savings,
            subset=(countries != "Japan"))
> summary(gj)
Coefficients:
              Estimate Std. Error t value Pr(>|t|)
(Intercept) 23.940171   7.783997    3.08   0.0036
pop15       -0.367901   0.153630   -2.39   0.0210
pop75       -0.973674   1.155450   -0.84   0.4040
dpi         -0.000471   0.000919   -0.51   0.6112
ddpi         0.334749   0.198446    1.69   0.0987
```

```
Residual standard error: 3.74 on 44 degrees of freedom
Multiple R-Squared: 0.277,        Adjusted R-squared: 0.211
F-statistic: 4.21 on 4 and 44 DF,  p-value: 0.00565
```

Comparing this to the full data fit, we observe several qualitative changes. Notice that the ddpi term is no longer significant and that the R^2 value has decreased a lot.

4.3 Checking the Structure of the Model

How do we check whether the systematic part ($Ey = X\beta$) of the model is correct? Lack of fit tests can be used when there is replication, which does not happen too often; however, even if you do have it, the tests do not tell you how to improve the model.

We can look at plots of $\hat{\varepsilon}$ against $\hat{y}$ and x_i to reveal problems or just simply look at plots of y against each x_i. The drawback to these plots is that the other predictors impact the relationship. *Partial regression* or *added variable* plots can help isolate the effect of x_i on y. Suppose we regress y on all x except x_i, and get residuals $\hat{\delta}$. These represent y with the other X-effect taken out. Similarly, if we regress x_i on all x except x_i, and get residuals $\hat{\gamma}$, we have the effect of x_i with the other X-effect taken out. The added variable plot shows $\hat{\delta}$ against $\hat{\gamma}$. Look for nonlinearity and outliers and/or influential observations in the plot.

The slope of a line fitted to the plot is $\hat{\beta}_i$. The partial regression plot provides some intuition about the meaning of regression coefficients. We are looking at the marginal relationship between the response and the predictor after the effect of the other predictors has been removed. Multiple regression is difficult because we cannot visualize the full relationship because of the high dimensionality. The partial regression plot allows us to focus on the relationship between one predictor and the response, much as in simple regression.

We illustrate using the savings dataset as an example again. We construct a partial regression (added variable) plot for pop15:

```
> d <- residuals(lm(sr ~ pop75 + dpi + ddpi,savings))
> m <- residuals(lm(pop15 ~ pop75 + dpi + ddpi,savings))
> plot(m,d,xlab="pop15 residuals",ylab="Savings residuals")
```

Compare the slope on the plot to the original regression and show the line on the plot (see Figure 4.13):

```
> coef(lm(d ~ m))
(Intercept)               m
 5.4259e-17 -4.6119e-01
> coef(g)
(Intercept)          pop15          pop75          dpi          ddpi
 28.5660865  -0.4611931  -1.6914977  -0.0003369    0.4096949
> abline(0,coef(g)['pop15'])
```

Notice how the slope in the plot and the slope for pop15 in the regression fit are the same.

Partial residual plots are a competitor to added variable plots. These plot $\hat{\varepsilon} + \hat{\beta}_i x_i$ against x_i. To see the motivation, look at the response with the predicted effect of the other X removed:

$$y - \sum_{j \neq i} x_j \hat{\beta}_j = \hat{y} + \hat{\varepsilon} - \sum_{j \neq i} x_j \hat{\beta}_j = x_i \hat{\beta}_i + \hat{\varepsilon}$$

Again the slope on the plot will be $\hat{\beta}_i$ and the interpretation is the same. Partial residual plots are reckoned to be better for nonlinearity detection while added variable plots are better for outlier/influential detection.

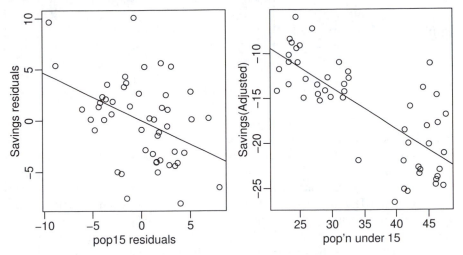

Figure 4.13 *Partial regression (left) and partial residual (right) plots for the savings data.*

A partial residual plot is easier to construct:

```
> plot(savings$pop15,residuals(g)+coef(g)['pop15']*savings$pop15,
  xlab="pop'n under 15", ylab="Savings(Adjusted)")
> abline(0,coef(g)['pop15'])
```

Or more directly using a function from the faraway package:

```
> prplot(g,1)
```

We see the two groups in the plot. It suggests that there may be a different relationship in the two groups. We investigate this:

```
> g1 <- lm(sr ~ pop15+pop75+dpi+ddpi,savings,subset=(pop15 > 35))
> g2 <- lm(sr ~ pop15+pop75+dpi+ddpi,savings,subset=(pop15 < 35))
> summary(g1)
Coefficients:
              Estimate Std. Error t value Pr(>|t|)
(Intercept) -2.433969  21.155028   -0.12     0.91
pop15        0.273854   0.439191    0.62     0.54
pop75       -3.548477   3.033281   -1.17     0.26
dpi          0.000421   0.005000    0.08     0.93
ddpi         0.395474   0.290101    1.36     0.19

Residual standard error: 4.45 on 18 degrees of freedom
Multiple R-Squared: 0.156,      Adjusted R-squared: -0.0319
F-statistic: 0.83 on 4 and 18 DF,  p-value: 0.523

> summary(g2)
Coefficients:
              Estimate Std. Error t value Pr(>|t|)
(Intercept) 23.961795   8.083750    2.96   0.0072
```

pop15	-0.385898	0.195369	-1.98	0.0609
pop75	-1.327742	0.926063	-1.43	0.1657
dpi	-0.000459	0.000724	-0.63	0.5326
ddpi	0.884394	0.295341	2.99	0.0067

Residual standard error: 2.77 on 22 degrees of freedom
Multiple R-Squared: 0.507,   Adjusted R-squared: 0.418
F-statistic: 5.66 on 4 and 22 DF,   p-value: 0.00273

In the first regression on the subset of underdeveloped countries, we find no relation between the predictors and the response. The p-value is 0.523. We know from
our previous examination of these data that this result is not attributable to outliers or
unsuspected transformations. In contrast, there is a strong relationship in the developed countries. The strongest predictor is growth with a suspicion of some relationship to proportion under 15. This latter effect has been reduced from prior analyses
because we have reduced the range of this predictor by the subsetting operation. The
graphical analysis has shown a relationship in the data that a purely numerical analysis might easily have missed.

Higher dimensional plots can also be useful for detecting structure that cannot be
seen in two dimensions. These are interactive in nature so you need to try them to
see how they work. We can make three-dimensional plots where color, point size and
rotation are used to give the illusion of a third dimension. We can also link two or
more plots so that points which are *brushed* in one plot are highlighted in another.

These tools look good but it is not clear whether they actually are useful in practice.
Certainly there are communication difficulties, as these plots cannot be easily printed.
R itself does not have such tools, but GGobi is a useful free tool for exploring higher
dimensional data that can be called from R. See www.ggobi.org.

Nongraphical techniques for checking the structural form of the model usually involve proposing alternative transformations or recombinations of the variables. This
approach is explored in the chapter on transformation.

Excercises

1. Using the sat dataset, fit a model with the total SAT score as the response and
   expend, salary, ratio and takers as predictors. Perform regression diagnostics on
   this model to answer the following questions. Display any plots that are relevent.
   Do not provide any plots about which you have nothing to say.

   (a) Check the constant variance assumption for the errors.
   (b) Check the normality assumption.
   (c) Check for large leverage points.
   (d) Check for outliers.
   (e) Check for influential points.
   (f) Check the structure of the relationship between the predictors and the response.

2. Using the teengamb dataset, fit a model with gamble as the response and the
   other variables as predictors. Answer the questions posed in the previous question.

3. For the `prostate` data, fit a model with `lpsa` as the response and the other variables as predictors. Answer the questions posed in the first question.

4. For the `swiss` data, fit a model with `Fertility` as the response and the other variables as predictors. Answer the questions posed in the first question.

5. For the `divusa` data, fit a model with `divorce` as the response and the other variables, except `year` as predictors. Check for serial correlation.

CHAPTER 5

Problems with the Predictors

5.1 Errors in the Predictors

The regression model $Y = X\beta + \varepsilon$ allows for Y being measured with error by having the ε term, but what if the X is measured with error? In other words, what if the X we see is not the X used to generate Y? It is not unreasonable that there might be errors in measuring X. For example, consider the problem of determining the effects of being exposed to a potentially hazardous substance such as secondhand tobacco smoke. Such exposure would be a predictor in such a study, but clearly it is very hard to measure this exactly over a period of years.

One should not confuse the errors in predictors with treating X as a random variable. For observational data, X could be regarded as a random variable, but the regression inference proceeds conditional on a fixed value for X. We make the assumption that the Y is generated conditional on the fixed value of X. Contrast this with the errors in predictors case where the X we see is not the X that was used to generate the Y.

Suppose that what we observe is (x_i^O, y_i^O) for $i = 1, \ldots n$ which are related to the true values (x_i^A, y_i^A):

$$
\begin{aligned}
y_i^O &= y_i^A + \varepsilon_i \\
x_i^O &= x_i^A + \delta_i
\end{aligned}
$$

where the errors ε and δ are independent. The situation is depicted in Figure 5.1. The true underlying relationship is:

$$ y_i^A = \beta_0 + \beta_1 x_i^A $$

but we only see (x_i^O, y_i^O). Putting it together, we get:

$$ y_i^O = \beta_0 + \beta_1 x_i^O + (\varepsilon_i - \beta_1 \delta_i) $$

Suppose we use least squares to estimate β_0 and β_1. Let's assume $E\varepsilon_i = E\delta_i = 0$ and that var $\varepsilon_i = \sigma_\varepsilon^2$, var $\delta_i = \sigma_\delta^2$. Let:

$$ \sigma_x^2 = \sum (x_i^A - \bar{x}^A)^2 / n \qquad \sigma_{x\delta} = cov(x^A, \delta) $$

For observational data, σ_x^2 is (almost) the sample variance of X^A while for a controlled experiment we can view it as just a numerical measure of the spread of the design. A similar distinction should be made for $cov(x^A, \delta)$ although in many cases, it will be reasonable to assume that this is zero.

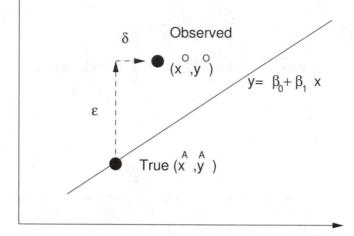

Figure 5.1 *Measurement error: True vs. observed data.*

Now $\hat{\beta}_1 = \sum(x_i - \bar{x})y_i / \sum(x_i - \bar{x})^2$ and after some calculation we find that:

$$E\hat{\beta}_1 = \beta_1 \frac{(\sigma_x^2 + \sigma_{x\delta})}{(\sigma_x^2 + \sigma_\delta^2 + 2\sigma_{x\delta})}$$

There are two main special cases of interest:

1. If there is no relation between X^A and δ, $\sigma_{x\delta} = 0$, this simplifies to:

$$E\hat{\beta}_1 = \beta_1 \frac{1}{1 + \sigma_\delta^2/\sigma_x^2}$$

So $\hat{\beta}_1$ will be biased towards zero, regardless of the sample size. If σ_δ^2 is small relative to σ_x^2, then the problem can be ignored. In other words, if the variability in the errors of observation of X are small relative to the range of X, then we need not be too concerned. For multiple predictors, the usual effect of measurement errors is also to bias the $\hat{\beta}$ in the direction of zero.

2. In controlled experiments, we need to distinguish two ways in which error in x may arise. In the first case, we measure x so although the true value is x^A, we observe x^0. If we were to repeat the measurement, we would have the same x^A, but a different x^0. In the second case, you fix x^0 — for example, you make up a chemical solution with a specified concentration x^0. The true concentration would be x^A. Now if you were to repeat this, you would get the same x^0, but the x^A would be different. In this latter case we have:

$$\sigma_{x\delta} = cov(X^0 - \delta, \delta) = -\sigma_\delta^2$$

and then we would have $E\hat{\beta}_1 = \beta_1$. So our estimate would be unbiased. This seems paradoxical, until you notice that the second case effectively reverses the roles of

x^A and x^O and if you get to observe the true X, then you will get an unbiased estimate of β_1. See Berkson (1950) for a discussion of this.

If the model is used for prediction purposes, we can make the same argument as in the second case above. In repeated "experiments," the value of x at which the prediction is to be made will be fixed, even though these may represent different underlying "true" values of x.

In cases where the error in X can simply not be ignored, we should consider alternatives to the least squares estimation of β. The least squares regression equation can be written as:

$$\frac{y - \bar{y}}{SD_y} = r\frac{(x - \bar{x})}{SD_x}$$

so that $\hat{\beta}_1 = rSD_y/SD_x$. Note that if we reverse the roles of x and y, we do not get the same regression equation. Since we have errors in both x and y in our problem, we might argue that neither one, in particular, deserves the role of response or predictor and so the equation should be the same either way. One way to achieve this is to set $\hat{\beta}_1 = SD_y/SD_x$. This is known as the *geometric mean functional relationship*. More on this can be found in Draper and Smith (1998). Another approach is to use the SIMEX method of Cook and Stefanski (1994), which we illustrate below.

Consider some data on the speed and stopping distances of cars in the 1920s. We plot the data, as seen in Figure 5.2, and fit a linear model:

```
> data(cars)
> plot(dist ~ speed, cars, ylab="distance")
> g <- lm(dist ~ speed, cars)
> summary(g)
Coefficients:
            Estimate Std. Error t value Pr(>|t|)
(Intercept)  -17.579      6.758   -2.60    0.012
speed          3.932      0.416    9.46  1.5e-12

Residual standard error: 15.4 on 48 degrees of freedom
Multiple R-Squared: 0.651,      Adjusted R-squared: 0.644
F-statistic: 89.6 on 1 and 48 DF,  p-value: 1.49e-12
> abline(g)
```

We could explore transformations and diagnostics for these data, but we will just focus on the measurement error issue. Now we investigate the effect of adding measurement error to the predictor. We plot the modified fits in Figure 5.2:

```
> ge1 <- lm(dist ~ I(speed+rnorm(50)), cars)
> coef(ge1)
          (Intercept) I(speed + rnorm(50))
             -15.0619               3.7582
> abline(ge1,lty=2)
> ge2 <- lm(dist ~ I(speed+2*rnorm(50)), cars)
> coef(ge2)
          (Intercept) I(speed + 2 * rnorm(50))
              -5.3503                   3.1676
> abline(ge2,lty=3)
```

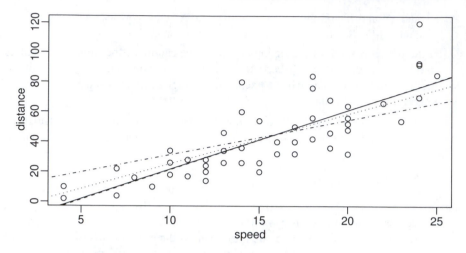

Figure 5.2 *Stopping distance and speeds of cars. The least squares fit is shown as a solid line. The fits with three progressively larger amounts of measurement error on the speed are shown as dotted lines, where the slope gets shallower as the error increases.*

```
> ge5 <- lm(dist ~ I(speed+5*rnorm(50)), cars)
> coef(ge5)
                (Intercept) I(speed + 5 * rnorm(50))
                    15.1589                   1.8696
> abline(ge5,lty=4)
```

We can see that the slope becomes shallower as the amount of noise increases.

Suppose we knew that the predictor, speed, in the original data had been measured with a known error variance, say 0.5. Given what we have seen in the simulated measurement error models, we might extrapolate back to suggest an estimate of the slope under no measurement error. This is the idea behind SIMEX.

Here we simulate the effects of adding normal random error with variances ranging from 0.1 to 0.5, replicating the experiment 1000 times for each setting:

```
> vv <- rep(1:5/10,each=1000)
> slopes <- numeric(5000)
> for(i in 1:5000) slopes[i] <- lm(dist ~
  I(speed+sqrt(vv[i])*rnorm(50)), cars)$coef[2]
```

Now plot the mean slopes for each variance. We are assuming that the data have variance 0.5 so the extra variance is added to this:

```
> betas <- c(coef(g)[2],colMeans(matrix(slopes,nrow=1000)))
> variances <- c(0,1:5/10)+0.5
> plot(variances,betas,xlim=c(0,1),ylim=c(3.86,4))
```

We fit a linear model and extrapolate to zero variance:

```
> gv <- lm(betas ~ variances)
> coef(gv)
```

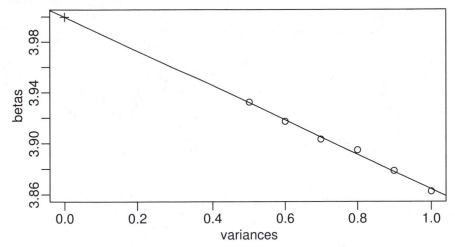

Figure 5.3 *Simulation–Extrapolation estimation of the unbiased slope in the presence of measurement error in the predictors. We predict* $\hat{\beta} = 4.0$ *at a variance of zero.*

```
(Intercept)     variances
    3.99975       -0.13552
> points(0,gv$coef[1],pch=3)
```

The predicted value of $\hat{\beta}$ at variance equal to zero, that is no measurement error, is 4.0. Better models for extrapolation are worth considering; see Cook and Stefanski (1994) for details.

5.2 Changes of Scale

Suppose we reexpress x_i as $(x_i + a)/b$. We might want to do this because predictors of similar magnitude result in $\hat{\beta}$ of similar sizes. $\hat{\beta} = 3.51$ is easier to parse than $\hat{\beta} = 0.000000351$ and we can choose a and b to achieve this. Furthermore, a change of units might aid interpretability. Finally, numerical stability in estimation is enhanced when all the predictors are on a similar scale.

Rescaling x_i leaves the t- and F-tests and $\hat{\sigma}^2$ and R^2 unchanged and $\hat{\beta}_i \rightarrow b\hat{\beta}_i$. Rescaling y in the same way leaves the t- and F-tests and R^2 unchanged, but $\hat{\sigma}$ and $\hat{\beta}$ will rescaled by b.

To demonstrate this, we use the same old model:

```
> data(savings)
> g <- lm(sr ~ pop15+pop75+dpi+ddpi,savings)
> summary(g)
Coefficients:
               Estimate Std. Error t value Pr(>|t|)
(Intercept) 28.566087   7.354516    3.88  0.00033
pop15       -0.461193   0.144642   -3.19  0.00260
pop75       -1.691498   1.083599   -1.56  0.12553
```

```
dpi            -0.000337   0.000931   -0.36  0.71917
ddpi            0.409695   0.196197    2.09  0.04247
```

```
Residual standard error: 3.8 on 45 degrees of freedom
Multiple R-Squared: 0.338,        Adjusted R-squared: 0.28
F-statistic: 5.76 on 4 and 45 DF,   p-value: 0.00079
```

The coefficient for income is rather small — let's measure income in thousands of dollars instead and refit:

```
> g <- lm(sr ~ pop15+pop75+I(dpi/1000)+ddpi,savings)
> summary(g)
Coefficients:
              Estimate Std. Error t value Pr(>|t|)
(Intercept)    28.566      7.355    3.88   0.00033
pop15          -0.461      0.145   -3.19   0.00260
pop75          -1.691      1.084   -1.56   0.12553
I(dpi/1000)    -0.337      0.931   -0.36   0.71917
ddpi            0.410      0.196    2.09   0.04247
```

```
Residual standard error: 3.8 on 45 degrees of freedom
Multiple R-Squared: 0.338,        Adjusted R-squared: 0.28
F-statistic: 5.76 on 4 and 45 DF,   p-value: 0.00079
```

What changed and what stayed the same?

One rather thorough approach to scaling is to convert all the variables to standard units (mean 0 and variance 1) using the scale() command:

```
> scsav <- data.frame(scale(savings))
> g <- lm(sr ~ ., scsav)
> summary(g)
Coefficients:
              Estimate Std. Error t value Pr(>|t|)
(Intercept)   4.0e-16     0.1200  3.3e-15   1.0000
pop15        -0.9420     0.2954   -3.19    0.0026
pop75        -0.4873     0.3122   -1.56    0.1255
dpi          -0.0745     0.2059   -0.36    0.7192
ddpi          0.2624     0.1257    2.09    0.0425
```

```
Residual standard error: 0.849 on 45 degrees of freedom
Multiple R-Squared: 0.338,        Adjusted R-squared: 0.28
F-statistic: 5.76 on 4 and 45 DF,   p-value: 0.00079
```

As may be seen, the intercept is zero. This is because the regression plane always runs through the point of the averages, which because of the centering, is now at the origin. Such scaling has the advantage of putting all the predictors and the response on a comparable scale, which makes comparisons simpler. It also allows the coefficients to be viewed as a kind of partial correlation — the values will always be between minus one and one. It also avoids some numerical problems that can arise when variables are of very different scales. The downside of this scaling is that the regression coefficients now represent the effect of a one standard unit increase in the predictor on the response in standard units — this might not always be easy to interpret.

5.3 Collinearity

When some predictors are linear combinations of others, then $X^T X$ is singular, and we have (exact) collinearity. There is no unique least squares estimate of β. If $X^T X$ is close to singular, we have collinearity (some call it multicollinearity). This causes serious problems with the estimation of β and associated quantities, as well as the interpretation. Collinearity can be detected in several ways:

1. Examination of the correlation matrix of the predictors may reveal large *pairwise* collinearities.

2. A regression of x_i on all other predictors gives R_i^2. Repeat for all predictors. R_i^2 close to one indicates a problem. The offending linear combination may be discovered by examining these regression coefficients.

3. Examine the eigenvalues of $X^T X$, where λ_1 is the largest eigenvalue with the others in decreasing order. Relatively small eigenvalues indicate a problem. The condition number is defined as:

$$\kappa = \sqrt{\frac{\lambda_1}{\lambda_p}}$$

where $\kappa \geq 30$ is considered large. κ is called the condition number. Other condition numbers, $\sqrt{\lambda_1/\lambda_i}$ are also worth considering because they indicate whether more than just one independent linear combination is to blame. Alternative calculations involve standardizing the predictors and/or excluding the intercept term.

Collinearity makes some of the parameters hard to estimate. Define:

$$S_{x_j x_j} = \sum_i (x_{ij} - \bar{x}_j)^2$$

then:

$$\text{var } \hat{\beta}_j = \sigma^2 \left(\frac{1}{1 - R_j^2}\right) \frac{1}{S_{x_j x_j}}$$

We can see that if x_j does not vary much, then the variance of $\hat{\beta}_j$ will be large. Another consequence of this equation is that it tells us which designs will minimize the variance of the regression coefficients if we have the ability to place the X. Orthogonality means that $R_j^2 = 0$ which minimizes the variance. Also we can maximize $S_{x_j x_j}$ by spreading X as much as possible. The maximum is attained by placing half the points at the minimum practical value and half at the maximum. Unfortunately, this design assumes the linearity of the effect and would make it impossible to check for any curvature. So, in practice, most would put some design points in the middle of the range to allow checking of the fit.

If R_j^2 is close to one, then the *variance inflation factor* $\frac{1}{1 - R_j^2}$ will be large.

Collinearity leads to imprecise estimates of β. The signs of the coefficients can be the opposite of what intuition about the effect of the predictor might suggest. The standard errors are inflated so that t-tests may fail to reveal significant factors. The fit becomes very sensitivite to measurement errors where small changes in y can lead to large changes in $\hat{\beta}$.

Car drivers like to adjust the seat position for their own comfort. Car designers would find it helpful to know where different drivers will position the seat depending on their size and age. Researchers at the HuMoSim laboratory at the University of Michigan collected data on 38 drivers. They measured age in years, weight in pounds, height with shoes and without shoes in cm, seated height arm length, thigh length, lower leg length and hipcenter the horizontal distance of the midpoint of the hips from a fixed location in the car in mm. We fit a model with all the predictors:

```
> data(seatpos)
> g <- lm(hipcenter ~ ., seatpos)
> summary(g)
Coefficients:
              Estimate Std. Error t value Pr(>|t|)
(Intercept) 436.4321    166.5716    2.62    0.014
Age           0.7757      0.5703    1.36    0.184
Weight        0.0263      0.3310    0.08    0.937
HtShoes      -2.6924      9.7530   -0.28    0.784
Ht            0.6013     10.1299    0.06    0.953
Seated        0.5338      3.7619    0.14    0.888
Arm          -1.3281      3.9002   -0.34    0.736
Thigh        -1.1431      2.6600   -0.43    0.671
Leg          -6.4390      4.7139   -1.37    0.182

Residual standard error: 37.7 on 29 degrees of freedom
Multiple R-Squared: 0.687,        Adjusted R-squared:   0.6
F-statistic: 7.94 on 8 and 29 DF,  p-value: 1.31e-05
```

This model already shows the signs of collinearity. The p-value for the F-statistics is very small and the R^2 is quite substantial, but none of the individual predictors is significant. We take a look at the pairwise correlations:

```
> round(cor(seatpos),3)
            Age Weight HtShoes     Ht Seated    Arm  Thigh
Age       1.000  0.081  -0.079 -0.090 -0.170  0.360  0.091
Weight    0.081  1.000   0.828  0.829  0.776  0.698  0.573
HtShoes  -0.079  0.828   1.000  0.998  0.930  0.752  0.725
Ht       -0.090  0.829   0.998  1.000  0.928  0.752  0.735
Seated   -0.170  0.776   0.930  0.928  1.000  0.625  0.607
Arm       0.360  0.698   0.752  0.752  0.625  1.000  0.671
Thigh     0.091  0.573   0.725  0.735  0.607  0.671  1.000
Leg      -0.042  0.784   0.908  0.910  0.812  0.754  0.650
hipcenter 0.205 -0.640  -0.797 -0.799 -0.731 -0.585 -0.591
            Leg hipcenter
Age      -0.042     0.205
Weight    0.784    -0.640
```

```
HtShoes       0.908      -0.797
Ht            0.910      -0.799
Seated        0.812      -0.731
Arm           0.754      -0.585
Thigh         0.650      -0.591
Leg           1.000      -0.787
hipcenter    -0.787       1.000
```

There are several large pairwise correlations both between predictors and between predictors and the response. Now we check the eigendecomposition:

```
> x <- model.matrix(g)[,-1]
> e <- eigen(t(x) %*% x)
> e$val
[1] 3.6537e+06 2.1479e+04 9.0432e+03 2.9895e+02 1.4839e+02
[6] 8.1174e+01 5.3362e+01 7.2982e+00
> sqrt(e$val[1]/e$val)
[1]    1.000   13.042   20.100  110.551  156.912  212.156  261.667
[8] 707.549
```

There is a wide range in the eigenvalues and several condition numbers are large. This means that problems are being caused by more than just one linear combination. Now check the variance inflation factors (VIFs). For the first variable this is:

```
> summary(lm(x[,1] ~ x[,-1]))$r.squared
[1] 0.49948
> 1/(1-0.49948)
[1] 1.9979
```

which is moderate in size — the VIF for orthogonal predictors is one. Now we compute all the VIFs in one go, using a function from the faraway package:

```
> vif(x)
     Age   Weight  HtShoes        Ht    Seated       Arm     Thigh
  1.9979   3.6470 307.4294  333.1378    8.9511    4.4964    2.7629
     Leg
  6.6943
```

There is much variance inflation. For example, we can interpret $\sqrt{307.4} = 17.5$ as telling us that the standard error for height with shoes is 17.5 times larger than it would have been without collinearity. We cannot apply this as a correction because we did not actually observe orthogonal data, but it does give us a sense of the size of the effect.

There is substantial instability in these estimates. Measuring the hipcenter is difficult to do accurately and we can expect some variation in these values. Suppose the measurement error had a SD of 10 mm. Let's see what happens when we add a random perturbation of this size to the response:

```
> g <- lm(hipcenter+10*rnorm(38) ~ ., seatpos)
> summary(g)
Coefficients:
              Estimate Std. Error t value Pr(>|t|)
(Intercept)    501.295    164.752    3.04   0.0049
```

```
Age                0.632      0.564     1.12    0.2720
Weight             0.129      0.327     0.39    0.6967
HtShoes           -2.329      9.647    -0.24    0.8109
Ht                 0.705     10.019     0.07    0.9444
Seated            -0.751      3.721    -0.20    0.8414
Arm               -0.951      3.858    -0.25    0.8070
Thigh             -1.879      2.631    -0.71    0.4808
Leg               -7.087      4.662    -1.52    0.1393
```

```
Residual standard error: 37.3 on 29 degrees of freedom
Multiple R-Squared: 0.696,        Adjusted R-squared: 0.612
F-statistic:   8.3 on 8 and 29 DF,  p-value: 8.71e-06
```

Although the R^2 and standard error are very similar to the previous fit, we see much larger changes in the coefficients indicating their sensitivity to the response values caused by the collinearity.

One cure for collinearity is amputation. We have too many variables that are trying to do the same job of explaining the response. When several variables, which are highly correlated, are each associated with the response, we have to take care that we do not conclude that the variables we drop have nothing to do with the response. Examine the full correlation matrix above. Consider just the correlations of the length variables:

```
> round(cor(x[,3:8]),2)
         HtShoes   Ht Seated  Arm Thigh   Leg
HtShoes     1.00 1.00   0.93 0.75  0.72  0.91
Ht          1.00 1.00   0.93 0.75  0.73  0.91
Seated      0.93 0.93   1.00 0.63  0.61  0.81
Arm         0.75 0.75   0.63 1.00  0.67  0.75
Thigh       0.72 0.73   0.61 0.67  1.00  0.65
Leg         0.91 0.91   0.81 0.75  0.65  1.00
```

These six variables are strongly correlated with each other — any one of them might do a good job of representing the other. We pick height as the simplest to measure. We are not claiming that the other predictors are not associated with the response, just that we do not need them all to predict the response:

```
> g2 <- lm(hipcenter ~ Age + Weight + Ht, seatpos)
> summary(g2)
Coefficients:
              Estimate Std. Error t value Pr(>|t|)
(Intercept) 528.29773  135.31295    3.90  0.00043
Age           0.51950    0.40804    1.27  0.21159
Weight        0.00427    0.31172    0.01  0.98915
Ht           -4.21190    0.99906   -4.22  0.00017
```

```
Residual standard error: 36.5 on 34 degrees of freedom
Multiple R-Squared: 0.656,        Adjusted R-squared: 0.626
F-statistic: 21.6 on 3 and 34 DF,  p-value: 5.13e-08
```

Comparing this with the original fit, we see that the fit is very similar in terms of R^2, but much fewer predictors are used. Further simplification is clearly possible.

If you must keep all your variables in the model, you should consider alternative methods of estimation such as ridge regression.

The effect of collinearity on prediction depends on where the prediction is to be made. The greater the distance is from the observed data, the more unstable the prediction. Distance needs to be considered in a Mahalanobis rather than a Euclidean sense.

Exercises

1. Using the `faithful` data, fit a regression of `duration` on `waiting`. Assuming that there was a measurement error in `waiting` of 30 seconds, use the SIMEX method to obtain a better estimate of the slope.

2. What would happen if the SIMEX method was applied to the response error variance rather than predictor measurement error variance?

3. Using the `divusa` data:

   (a) Fit a regression model with `divorce` as the response and `unemployed`, `femlab`, `marriage`, `birth` and `military` as predictors. Compute the condition numbers and interpret their meaning.

   (b) For the same model, compute the VIFs. Is there evidence that collinearity causes some predictors not to be significant? Explain.

   (c) Does the removal of insignificant predictors from the model reduce the collinearity? Investigate.

4. For the `longley` data, fit a model with `Employed` as the response and the other variables as predictors.

   (a) Compute and comment on the condition numbers.

   (b) Compute and comment on the correlations between the predictors.

   (c) Compute the variance inflation factors.

5. For the `prostate` data, fit a model with `lpsa` as the response and the other variables as predictors.

   (a) Compute and comment on the condition numbers.

   (b) Compute and comment on the correlations between the predictors.

   (c) Compute the variance inflation factors.

CHAPTER 6

Problems with the Error

The standard assumption about the error term ε is that it is independent and identically distributed (i.i.d.) from case to case. That is, var $\varepsilon = \sigma^2 I$. Furthermore, we also assume that the errors are normally distributed in order to carry out the usual statistical inference. We have seen that these assumptions can often be violated and we must then consider alternatives. When the errors are not i.i.d., we consider the use of *generalized least squares* (GLS). When the errors are independent, but not identically distributed, we can use *weighted least squares* (WLS), which is a special case of GLS. Sometimes, we have a good idea how large the error should be, but the residuals may be much larger than we expect. This is evidence of a *lack of fit*. When the errors are not normally distributed, we can use *robust regression*.

6.1 Generalized Least Squares

Until now we have assumed that var $\varepsilon = \sigma^2 I$, but sometimes the errors have non-constant variance or are correlated. Suppose instead that var $\varepsilon = \sigma^2 \Sigma$ where σ^2 is unknown but Σ is known — in other words, we know the correlation and relative variance between the errors, but we do not know the absolute scale. Right now, it might seem redundant to distinguish between σ and Σ, but we will see how this will be useful later.

We can write $\Sigma = SS^T$, where S is a triangular matrix using the Choleski decomposition. Now we can transform the regression model as follows:

$$
\begin{aligned}
y &= X\beta + \varepsilon \\
S^{-1}y &= S^{-1}X\beta + S^{-1}\varepsilon \\
y' &= X'\beta + \varepsilon'
\end{aligned}
$$

Now we find that:

$$\text{var } \varepsilon' = \text{var } (S^{-1}\varepsilon) = S^{-1}(\text{var } \varepsilon)S^{-T} = S^{-1}\sigma^2 SS^T S^{-T} = \sigma^2 I$$

So we can reduce GLS to ordinary least squares (OLS) by a regression of $y' = S^{-1}y$ on $S^{-1}X$ which has error ε' that is i.i.d. So we simply reduce the problem to one that we have already solved. In this transformed model, the sum of squares is:

$$(S^{-1}y - S^{-1}X\beta)^T(S^{-1}y - S^{-1}X\beta) = (y - X\beta)^T S^{-T} S^{-1}(y - X\beta) = (y - X\beta)^T \Sigma^{-1}(y - X\beta)$$

which is minimized by:

$$\hat{\beta} = (X^T \Sigma^{-1} X)^{-1} X^T \Sigma^{-1} y$$

We find that:

$$\text{var } \hat{\beta} = (X^T \Sigma^{-1} X)^{-1} \sigma^2$$

Since $\varepsilon' = S^{-1}\varepsilon$, diagnostics should be applied to the residuals, $S^{-1}\hat{\varepsilon}$. If we have the right Σ, then these should be approximately i.i.d.

The main problem in applying GLS in practice is that Σ may not be known and we may have to estimate it. To illustrate this we will use a built-in R dataset known as Longley's regression data. Our response is the number of people employed, yearly from 1947 to 1962 and the predictors are gross national product (GNP) and population 14 years of age and over. The data originally appeared in Longley (1967).

Fit a linear model:

```
> data(longley)
> g <- lm(Employed ~ GNP + Population, longley)
> summary(g,cor=T)
Coefficients:
            Estimate Std. Error t value Pr(>|t|)
(Intercept)  88.9388    13.7850    6.45  2.2e-05
GNP           0.0632     0.0106    5.93  5.0e-05
Population   -0.4097     0.1521   -2.69    0.018

Residual standard error: 0.546 on 13 degrees of freedom
Multiple R-Squared: 0.979,         Adjusted R-squared: 0.976
F-statistic:  304 on 2 and 13 DF,  p-value: 1.22e-11

Correlation of Coefficients:
            (Intercept) GNP
GNP          0.98
Population  -1.00       -0.99
```

The correlation between the coefficients for GNP and Population is strongly negative while the correlation between the corresponding variables:

```
> cor(longley$GNP,longley$Pop)
[1] 0.99109
```

is strongly positive.

In data collected over time such as this, successive errors could be correlated. The simplest way to model this is the autoregressive form:

$$\varepsilon_{i+1} = \rho\varepsilon_i + \delta_i$$

where $\delta_i \sim N(0,\tau^2)$. We can estimate this correlation ρ by:

```
> cor(residuals(g)[-1],residuals(g)[-16])
[1] 0.31041
```

Under this assumption $\Sigma_{ij} = \rho^{|i-j|}$. For simplicity, let's assume we know that $\rho = 0.31041$. We now construct the Σ matrix and compute the GLS estimate of β along with its standard errors. The calculation is for demonstration purposes only:

```
> x <- model.matrix(g)
> Sigma <- diag(16)
> Sigma <- 0.31041^abs(row(Sigma)-col(Sigma))
> Sigi <- solve(Sigma)
> xtxi <- solve(t(x) %*% Sigi %*% x)
> (beta <- solve(t(x) %*% Sigi %*% x,t(x) %*%
```

```
        Sigi %*% longley$Empl))
                       [,1]
    (Intercept) 94.89889
    GNP          0.06739
    Population  -0.47427
> res <- longley$Empl - x %*% beta
> (sig <- sqrt((t(res) %*% Sigi %*% res)/g$df))
            [,1]
    [1,] 0.5424432
> sqrt(diag(xtxi))*sig
[1] 13.94477260  0.01070339  0.15338547
```

Compare with the model output above where the errors are assumed to be uncorrelated. Another way to get the same result is to regress $S^{-1}y$ on $S^{-1}x$ as we demonstrate here:

```
> sm <- chol(Sigma)
> smi <- solve(t(sm))
> sx <- smi %*% x
> sy <- smi %*% longley$Empl
> summary(lm(sy ~ sx-1))
Coefficients:
                 Estimate Std. Error t value Pr(>|t|)
sx(Intercept)     94.8989    13.9448    6.81  1.3e-05
sxGNP              0.0674     0.0107    6.30  2.8e-05
sxPopulation      -0.4743     0.1534   -3.09   0.0086

Residual standard error: 0.542 on 13 degrees of freedom
```

In practice, we would not know that the $\rho = 0.31$ and we will need to estimate it from the data. Our initial estimate is 0.31, but once we fit our GLS model we would need to reestimate it as:

```
> cor(res[-1],res[-16])
[1] 0.35642
```

and then recompute the model again with $\rho = 0.35642$. This process would be iterated until convergence. This is cumbersome. A more convenient approach may be found in the nlme package of Pinheiro and Bates (2000), which contains a GLS fitting function. We can use it to fit this model:

```
> library(nlme)
> g <- gls(Employed ~ GNP + Population,
  correlation=corAR1(form= ~Year), data=longley)
> summary(g)
Correlation Structure: AR(1)
 Formula: ~Year
 Parameter estimate(s):
    Phi
0.64417

Coefficients:
                Value Std.Error t-value p-value
```

```
(Intercept) 101.858    14.1989   7.1736   <.0001
GNP           0.072     0.0106   6.7955   <.0001
Population   -0.549     0.1541  -3.5588   0.0035
```

```
Residual standard error: 0.68921
Degrees of freedom: 16 total; 13 residual
```

We see that the estimated value of ρ is 0.64. However, if we check the confidence intervals for this:

```
> intervals(g)
Approximate 95% confidence intervals
```

```
 Coefficients:
                  lower        est.        upper
(Intercept) 71.183204 101.858133  132.533061
GNP          0.049159   0.072071    0.094983
Population  -0.881491  -0.548513   -0.215536
```

```
 Correlation structure:
        lower      est.    upper
Phi -0.44335 0.64417 0.96451
```

```
 Residual standard error:
   lower      est.     upper
0.24772 0.68921 1.91748
```

we see from the interval, $(-0.44, 0.96)$, that it is not significantly different from zero. So there is no evidence of serial correlation.

6.2 Weighted Least Squares

Sometimes the errors are uncorrelated, but have unequal variance where the form of the inequality is known. When Σ is diagonal, the errors are uncorrelated but do not necessarily have equal variance. WLS can be used in this situation. We can write $\Sigma = \text{diag}(1/w_1, \ldots, 1/w_n)$, where the w_i are the *weights* so $S = \text{diag}(\sqrt{1/w_1}, \ldots, \sqrt{1/w_n})$. So we can regress $\sqrt{w_i}y_i$ on $\sqrt{w_i}x_i$ (although the column of ones in the X-matrix needs to be replaced with $\sqrt{w_i}$). Cases with low variability should get a high weight, high variability a low weight. Some examples:

1. Errors proportional to a predictor: var $(\varepsilon_i) \propto x_i$ suggests $w_i = x_i^{-1}$.

2. When the Y_i are the averages of n_i observations, then var $y_i = $ var $\varepsilon_i = \sigma^2/n_i$, which suggests $w_i = n_i$. Responses that are averages arise quite commonly, but take care that the variance in the response really is proportional to the group size. For example, consider the life expectancy for different countries. At first glance, one might consider setting the weights equal to the populations of the countries, but notice that there are many other sources of variation in life expectancy that would dwarf the population size effect. Setting $w_i = n_i$ is only likely to be sensible for small n_i.

When weights are used, the residuals must be modified. Use $\sqrt{w_i}\hat{\varepsilon}_i$ for diagnostics.

Elections for the French presidency proceed in two rounds. In 1981, there were 10 candidates in the first round. The top two candidates then went on to the second round, which was won by François Mitterand over Valéry Giscard-d'Estaing. The losers in the first round can gain political favors by urging their supporters to vote for one of the two finalists. Since voting is private, we cannot know how these votes were transferred; we might hope to infer from the published vote totals how this might have happened. Anderson and Loynes (1987) published data on these vote totals in every fourth department of France:

```
> data(fpe)
> fpe
          EI    A    B    C    D    E    F    G    H    J    K   A2   B2
Ain      260   51   64   36   23    9    5    4    4    3    3  105  114
Alpes     75   14   17    9    9    3    1    2    1    1    1   32   31
...
```

A and B stand for Mitterand's and Giscard's votes in the first round, respectively, while A2 and B2 represent their votes in the second round. C−K are the first round votes of the other candidates while EI is *electeur inscrits* or registered voters. All numbers are in thousands. The total number of voters in the second round was greater than the first — we can compute the difference as N.

We will treat this group effectively as another first round candidate (we could reasonably handle this differently). Now we can represent the transfer of votes as:

$$A2 = \beta_A A + \beta_B B + \beta_C C + \beta_D D + \beta_E E + \beta_F F + \beta_G G + \beta_H H + \beta_J J + \beta_K K + \beta_N N$$

where β_i represents the proportion of votes transferred from candidate i to Mitterand in the second round. Now we would expect these transfer proportions to vary somewhat between departments, so if we treat the above as a regression equation, there will be some error from department to department. The error will have a variance in proportion to the number of voters because it will be like a variance of a sum rather than a mean. Since the weights should be inversely proportional to the variance, this suggests that the weights should be set to $1/EI$. Notice also that the equation has no intercept, hence the -1 in the model formula. We fit the appropriate model:

```
> g <- lm(A2 ~ A+B+C+D+E+F+G+H+J+K+N-1, fpe, weights=1/EI)
> coef(g)
        A         B         C         D         E         F         G
  1.06713  -0.10505   0.24596   0.92619   0.24940   0.75511   1.97221
        H         J         K         N
 -0.56622   0.61164   1.21066   0.52935
```

Note that the weights do matter — see what happens when we leave them out:

```
> lm(A2 ~ A+B+C+D+E+F+G+H+J+K+N-1, fpe)$coef
        A         B         C         D         E         F         G
  1.07515  -0.12456   0.25745   0.90454   0.67068   0.78253   2.16566
        H         J         K         N
 -0.85429   0.14442   0.51813   0.55827
```

which causes substantial changes for some of the lesser candidates. Furthermore, only the relative proportions of the weights matter — for example, suppose we multiply the weights by 53:

```
> lm(A2 ~ A+B+C+D+E+F+G+H+J+K+N-1, fpe, weights=53/EI)$coef
      A         B         C         D         E         F         G
 1.06713  -0.10505   0.24596   0.92619   0.24940   0.75511   1.97221
      H         J         K         N
-0.56622   0.61164   1.21066   0.52935
```

This makes no difference.

Now there is one remaining difficulty, unrelated to the weighting, in that proportions are supposed to be between zero and one. We can impose an *ad hoc* fix by truncating the coefficients that violate this restriction either to zero or one as appropriate. This gives:

```
> lm(A2 ~ offset(A+G+K)+C+D+E+F+N-1, fpe, weights=1/EI)$coef
      C         D         E         F         N
0.22577   0.96998   0.39020   0.74424   0.60854
```

We see that voters for the Communist candidate D apparently almost all voted for the Socialist Mitterand in the second round. However, we see that around 20% of the voters for the Gaullist candidate C voted for Mitterand. This is surprising since these voters would normally favor the more right wing candidate, Giscard. This appears to be the decisive factor. We see that of the larger blocks of smaller candidates, the Ecology party voters, E, roughly split their votes as did the first round nonvoters. The other candidates had very few voters and so their behavior is less interesting.

This analysis is somewhat crude and more sophisticated approaches are discussed in Anderson and Loynes (1987).

In cases where the form of the variance of ε is not completely known, we may model Σ using a small number of parameters. For example:

$$\text{var } \varepsilon_i = \gamma_0 + \gamma_1 x_1$$

might seem reasonable in a given situation. The iteratively reweighted least squares (IRWLS) fitting algorithm is:

1. Start with $w_i = 1$.

2. Use least squares to estimate β.

3. Use the residuals to estimate γ, perhaps by regressing $\hat{\varepsilon}^2$ on x.

4. Recompute the weights and go to 2.

Continue until convergence. There are some concerns about this because the estimation of the γ has some uncertainty and consumes some degrees of freedom. This affects the subsequent inference about β. An extensive investigation of this may be found in Carroll and Ruppert (1988).

Another approach is to model the variance and jointly estimate the regression and weighting parameters using a likelihood-based method. This can be implemented in R using the `gls()` function in the `nlme` library.

6.3 Testing for Lack of Fit

How can we tell whether a model fits the data? If the model is correct, then $\hat{\sigma}^2$ should be an unbiased estimate of σ^2. If we have a model that is not complex enough to fit the data or simply takes the wrong form, then $\hat{\sigma}^2$ will overestimate σ^2. The situation is illustrated in Figure 6.1. Alternatively, if our model is too complex and overfits the data, then $\hat{\sigma}^2$ will be an underestimate.

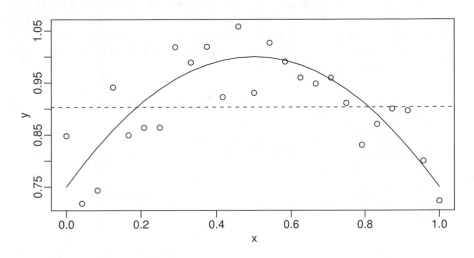

Figure 6.1 *True quadratic fit shown with the solid line and incorrect linear fit shown with the dotted line. Estimate of σ^2 will be unbiased for the quadratic model, but far too large for the linear model.*

This suggests a possible testing procedure — we should compare $\hat{\sigma}^2$ to σ^2. The usual problem, of course, is that we do not know the true value of σ^2 and so the comparison cannot be made. In a few cases, we might actually know σ^2 – for example, when measurement error is the only source of variation and we know its variance because we are very familiar with the measurement device. This is rather uncommon and we do not discuss it here — see Weisberg (1985) for an example. A more realistic possibility is that we have *replication* in our data that allows an estimate of σ^2 that does not depend on any particular model.

The $\hat{\sigma}^2$ that is based in the chosen regression model needs to be compared to some model-free estimate of σ^2. We can do this if we have repeated y for one or more fixed x. These replicates do need to be truly independent. They cannot just be repeated measurements on the same subject or unit. Such repeated measures would only reveal the within subject variability or the measurement error. We need to know the between subject variability, as this reflects the σ^2 described in the model. Let y_{ij} be the ith observation in the group of replicates j.

The "pure error" estimate of σ^2 is given by SS_{pe}/df_{pe} where:

$$SS_{pe} = \sum_j \sum_i (y_{ij} - \bar{y}_j)^2$$

and degrees of freedom $df_{pe} = \sum_j (\#replicates - 1) = n - \#groups.$

If you fit a model that assigns one parameter to each group of observations with

fixed x, then the $\hat{\sigma}^2$ from this model will be the pure error $\hat{\sigma}^2$. This model is just
the one-way analyis of variance (ANOVA) model — see Chapter 14. Comparing this
model to the regression model amounts to the lack of fit test.

The data for this example consist of 13 specimens of 90/10 Cu–Ni alloys with
varying percentages of iron content. The specimens were submerged in seawater for
60 days and the weight loss due to corrosion was recorded in units of milligrams per
square decimeter per day. The data come from Draper and Smith (1998). We load in
and plot the data, as seen in Figure 6.2:

```
> data(corrosion)
> plot(loss ~ Fe, corrosion,xlab="Iron content",
  ylab="Weight loss")
```

We fit a straight-line model:

```
> g <- lm(loss ~ Fe, corrosion)
> summary(g)
Coefficients:
              Estimate Std. Error t value Pr(>|t|)
(Intercept)     129.79       1.40    92.5  < 2e-16
Fe              -24.02       1.28   -18.8  1.1e-09

Residual standard error: 3.06 on 11 degrees of freedom
Multiple R-Squared: 0.97,        Adjusted R-squared: 0.967
F-statistic:  352 on 1 and 11 DF,  p-value: 1.06e-09
```

Now show the regression line on the plot:

```
> abline(coef(g))
```

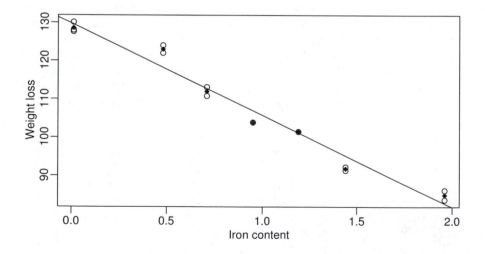

Figure 6.2 *Linear fit to the Cu–Ni corrosion data. Group means denoted by black diamonds.*

We have an R^2 of 97% and an apparently good fit to the data. We now fit a model
that reserves a parameter for each group of data with the same value of x. This is

accomplished by declaring the predictor to be a factor. We will describe this in more detail in Chapter 14:

```
> ga <- lm(loss ~ factor(Fe), corrosion)
```

The fitted values are the means in each group — put these on the plot:

```
> points(corrosion$Fe,fitted(ga),pch=18)
```

We can now compare the two models in the usual way:

```
> anova(g,ga)
Analysis of Variance Table

Model 1: loss ~ Fe
Model 2: loss ~ factor(Fe)
  Res.Df Res.Sum Sq Df Sum Sq F value Pr(>F)
1     11      102.9
2      6       11.8  5   91.1    9.28 0.0086
```

The low p-value indicates that we must conclude that there is a lack of fit. The reason is that the pure error SD $\sqrt{(11.8/6)} = 1.4$, is substantially less than the regression standard error of 3.06. We might investigate models other than a straight line although no obvious alternative is suggested by the plot. Before considering other models, we would first find out whether the replicates are genuine. Perhaps the low pure error SD can be explained by some correlation in the measurements. They may not be genuine replicates. Another possible explanation is that an unmeasured third variable is causing the lack of fit.

When there are replicates, it is impossible to get a perfect fit. Even when there is a parameter assigned to each group of x-values, the residual sum of squares will not be zero. For the factor model above, the R^2 is 99.7%. So even this saturated model does not attain a 100% value for R^2. For these data, it is a small difference but in other cases, the difference can be substantial. In these cases, one should realize that the maximum R^2 that may be attained might be substantially less than 100% and so perceptions about what a good value for R^2 should be downgraded appropriately.

These methods are good for detecting lack of fit, but if the null hypothesis is accepted, we cannot conclude that we have the true model. After all, it may be that we just did not have enough data to detect the inadequacies of the model. All we can say is that the model is not contradicted by the data.

When there are no replicates, it may be possible to group the responses for similar x, but this is not straightforward. It is also possible to detect lack of fit by less formal, graphical methods.

A more general question is how good a fit do you really want? By increasing the complexity of the model, it is possible to fit the data more closely. By using as many parameters as data points, we can fit the data exactly. Very little is achieved by doing this since we learn nothing beyond the data and any predictions made using such a model will tend to have a very high variance. The question of how complex a model to fit is difficult and fundamental. For example, we can fit the mean responses for the previous example exactly using a sixth order polynomial:

```
> gp <- lm(loss ~ Fe+I(Fe^2)+I(Fe^3)+I(Fe^4)+I(Fe^5)+
  I(Fe^6),corrosion)
```

Now look at this fit:

```
> plot(loss ~ Fe, data=corrosion,ylim=c(60,130))
> points(corrosion$Fe,fitted(ga),pch=18)
> grid <- seq(0,2,len=50)
> lines(grid,predict(gp,data.frame(Fe=grid)))
```

as shown in Figure 6.3. The fit of this model is excellent — for example:

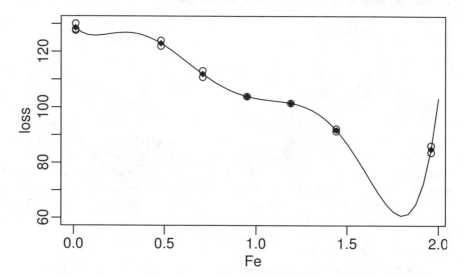

Figure 6.3 *Polynomial fit to the corrosion data.*

```
> summary(gp)$r.squared
[1] 0.99653
```

but it is clearly ridiculous. There is no plausible reason corrosion loss should suddenly drop at 1.7 and thereafter increase rapidly. This is a consequence of overfitting the data. This illustrates the need not to become too focused on measures of fit like R^2. The fit needs to reflect knowledge of the subject matter and simplicity in modeling is a virtue.

6.4 Robust Regression

When the errors are normal, least squares regression is clearly best, but when the errors are nonnormal, other methods may be considered. Of particular concern are long-tailed error distributions. One approach is to remove the largest residuals as outliers and still use least squares, but this may not be effective when there are several large residuals because of the leave-out-one nature of the outlier tests. Furthermore, the outlier test is an accept/reject procedure that is not smooth and may not be statistically efficient for the estimation of β. Robust regression provides an alternative. There are several methods, but we will present just two popular methods.

6.4.1 M-Estimation

M-estimates choose β to minimize:

$$\sum_{i=1}^{n} \rho(y_i - x_i^T \beta)$$

Some possible choices for ρ are:

1. $\rho(x) = x^2$ is just least squares.
2. $\rho(x) = |x|$ is called least absolute deviation (LAD) regression. This is also called L_1 regression.
3.

$$\rho(x) = \begin{cases} x^2/2 & \text{if } |x| \leq c \\ c|x| - c^2/2 & \text{otherwise} \end{cases}$$

is called Huber's method and is a compromise between least squares and LAD regression. c should be a robust estimate of σ. A value proportional to the median of $|\hat{\epsilon}|$ is suitable.

Robust regression is related to WLS. The normal equations tell us that:

$$X^T(y - X\hat{\beta}) = 0$$

With weights and in nonmatrix form this becomes:

$$\sum_{i=1}^{n} w_i x_{ij}(y_i - \sum_{j=1}^{p} x_{ij}\beta_j) = 0 \quad j = 1, \ldots p$$

Now differentiating the M-estimate criterion with respect to β_j and setting to zero we get:

$$\sum_{i=1}^{n} \rho'(y_i - \sum_{j=1}^{p} x_{ij}\beta_j)x_{ij} = 0 \quad j = 1, \ldots p$$

Now let $u_i = y_i - \sum_{j=1}^{p} x_{ij}\beta_j$ to get:

$$\sum_{i=1}^{n} \frac{\rho'(u_i)}{u_i} x_{ij}(y_i - \sum_{j=1}^{p} x_{ij}\beta_j) = 0 \quad j = 1, \ldots p$$

so we can make the identification of $w(u) = \rho'(u)/u$. We find for our choices of ρ above that:

1. LS: $w(u)$ is constant.
2. LAD: $w(u) = 1/|u|$ — note the asymptote at 0 — this makes a weighting approach infeasible.
3. Huber:

$$w(u) = \begin{cases} 1 & \text{if } |u| \leq c \\ c/|u| & \text{otherwise} \end{cases}$$

There are many other choices that have been used. Because the weights depend on the residuals, an IRWLS approach to fitting must be used. We can get standard errors by $\text{vâr } \hat{\beta} = \hat{\sigma}^2 (X^T W X)^{-1}$ using a robust estimate of σ^2.

We demonstrate the methods on the Galápagos Islands data. Using least squares first:

```
> data(gala)
> gl <- lm(Species ~ Area + Elevation + Nearest +
  Scruz + Adjacent,gala)
> summary(gl)
Coefficients:
             Estimate Std. Error t value Pr(>|t|)
(Intercept)   7.06822   19.15420    0.37   0.7154
Area         -0.02394    0.02242   -1.07   0.2963
Elevation     0.31946    0.05366    5.95  3.8e-06
Nearest       0.00914    1.05414    0.01   0.9932
Scruz        -0.24052    0.21540   -1.12   0.2752
Adjacent     -0.07480    0.01770   -4.23   0.0003

Residual standard error: 61 on 24 degrees of freedom
Multiple R-Squared: 0.766,        Adjusted R-squared: 0.717
F-statistic: 15.7 on 5 and 24 DF,   p-value: 6.84e-07
```

Least squares works well when there are normal errors, but performs poorly for long-tailed errors. The Huber method is the default choice of the rlm() function, which is part of the MASS package of Venables and Ripley (2002).

```
> library(MASS)
> gr <- rlm(Species ~ Area + Elevation + Nearest +
  Scruz + Adjacent,gala)
> summary(gr)
Coefficients:
             Value  Std. Error t value
(Intercept)  6.361  12.390        0.513
Area        -0.006   0.015       -0.421
Elevation    0.248   0.035        7.132
Nearest      0.359   0.682        0.527
Scruz       -0.195   0.139       -1.401
Adjacent    -0.055   0.011       -4.765

Residual standard error: 29.7 on 24 degrees of freedom
```

The R^2 and F-statistics are not given because they cannot be calculated (at least not in the same way). Similarly, p-values are not given although we can use the asymptotic normality of the estimator to make approximate inferences using the t-values. The numerical values of the coefficients have changed a small amount, but the general significance of the variables remains the same and our substantive conclusion would not be altered.

We can also do LAD regression using the quantreg package. The default option does LAD while other options allow for quantile regression:

```
> library(quantreg)
> attach(gala)
> gq <- rq(Species ~Area+Elevation+Nearest+Scruz+Adjacent)
> summary(gq)
```

```
Coefficients:
             coefficients    lower bd   upper bd
(Intercept)     1.3144484  -19.877769  24.374115
Area           -0.0030600   -0.031851   0.527999
Elevation       0.2321147    0.124526   0.501955
Nearest         0.1636558   -3.163385   2.988965
Scruz          -0.1231408   -0.479867   0.134763
Adjacent       -0.0518535   -0.104577   0.017394
```

```
Degrees of freedom: 30 total; 24 residual
> detach(gala)
```

Again, there is some change in the coefficients. The confidence intervals now suggest that adjacent is not significant.

For this example, we do not see any big qualitative difference in the coefficients and for want of evidence to the contrary, we might stick with least squares as the easiest to work with. Had we seen something different, we would need to find out the cause. Perhaps some group of observations were not being fit well and the robust regression excluded these points.

6.4.2 Least Trimmed Squares

Another popular method is least trimmed squares (LTS). Here one minimizes $\sum_{i=1}^{q} \hat{\varepsilon}_{(i)}^2$ where q is some number less than n and (i) indicates sorting. This method has a high *breakdown* point because it can tolerate a large number of outliers depending on how q is chosen. The Huber and L_1 methods will still fail if some $\varepsilon_i \to \infty$. LTS is an example of a *resistant* regression method. Resistant methods are good for dealing with data where we expect there to be a certain number of bad observations that we want to have no weight in the analysis:

```
> library(lqs)
> g <- ltsreg(Species ~ Area + Elevation + Nearest +
  Scruz + Adjacent,gala)
> coef(g)
(Intercept)        Area   Elevation      Nearest        Scruz
   7.410175    1.627483    0.011830     1.095214    -0.125413
   Adjacent
  -0.204292
> g <- ltsreg(Species ~ Area + Elevation + Nearest +
  Scruz + Adjacent,gala)
> coef(g)
(Intercept)        Area   Elevation      Nearest        Scruz
   5.972165    1.578810    0.025237     0.948129    -0.108829
   Adjacent
  -0.201755
```

The default choice of q is $\lfloor n/2 \rfloor + \lfloor (p+1)/2 \rfloor$ where $\lfloor x \rfloor$ indicates the largest integer less than or equal to x. I repeated the command twice and you will notice that the results are somewhat different. This is because the default genetic algorithm used to compute the coefficients is nondeterministic. An exhaustive search method can be used:

```
> g <- ltsreg(Species ~ Area + Elevation + Nearest + Scruz
  + Adjacent, gala,nsamp="exact")
> coef(g)
(Intercept)         Area    Elevation       Nearest         Scruz
   9.381145     1.543658     0.024125      0.811109     -0.117732
 Adjacent
-0.197923
```

This takes about 17 seconds on a 2.4 GHz Intel Pentium IV processor. For larger datasets, it will take much longer so this method might be impractical and the default method might be required.

This really does make substantial differences to the coefficients. For example, the Area coefficient is now substantially larger while the Elevation coefficient is substantially smaller. However, we do not have the standard errors for the LTS regression coefficients. We now use a general method for inference that is especially useful when such theory is lacking — the bootstrap.

To understand how this method works, think about how we might empirically determine the distribution of an estimator. We could repeatedly generate artificial data from the true model, compute the estimate each time and gather the results to study the distribution. This technique, called simulation, is not available to us for real data, because we do not know the true model. The bootstrap emulates the simulation procedure above except instead of sampling from the true model, it samples from the observed data. Remarkably, this technique is often effective. It sidesteps the need for theoretical calculations that may be extremely difficult or even impossible. See Efron and Tibshirani (1993) for an introductory text. To see how the bootstrap method compares with simulation, we spell out the steps involved. In both cases, we consider X fixed.

Simulation
The idea is to sample from the known distribution and compute the estimate, repeating many times to find as good an estimate of the sampling distribution of the estimator as we need. For the regression case, it is easiest to start with a sample from the error distribution since these are assumed to be independent and identically distributed:

1. Generate ε from the known error distribution.

2. Form $y = X\beta + \varepsilon$ from the known β and fixed X.

3. Compute $\hat{\beta}$.

We repeat these three steps many times. We can estimate the sampling distribution of $\hat{\beta}$ using the empirical distribution of the generated $\hat{\beta}$, which we can estimate as accurately as we please by simply running the simulation for long enough. This technique is useful for a theoretical investigation of the properties of a proposed new estimator. We can see how its performance compares to other estimators. However, it is of no value for the actual data since we do not know the true error distribution and we do not know the true β.

Bootstrap

The bootstrap method mirrors the simulation method, but uses quantities we do know. Instead of sampling from the population distribution, which we do not know in practice, we resample from the data:

1. Generate ε^* by sampling with replacement from $\hat{\varepsilon}_1, \ldots, \hat{\varepsilon}_n$.
2. Form $y^* = X\hat{\beta} + \varepsilon^*$.
3. Compute $\hat{\beta}^*$ from (X, y^*).

This time, we use only quantities that we know. For very small n, it is possible to compute $\hat{\beta}^*$ for every possible sample from $\hat{\varepsilon}_1, \ldots, \hat{\varepsilon}_n$, but usually we can only take as many samples as we have computing power available. This number of bootstrap samples can be as small as 50 if all we want is an estimate of the variance of our estimates but needs to be larger if confidence intervals are wanted.

To implement this, we need to be able to take a sample of residuals with replacement. `sample()` is good for generating random samples of indices:

```
> sample(10,rep=T)
 [1] 7 9 9 2 5 7 4 1 8 9
```

and hence a random sample (with replacement) of LTS residuals is:

```
> residuals(g)[sample(30,rep=TRUE)]
      Onslow      Seymour       Onslow       Rabida Daphne.Major
   -5.252656    33.841670    -5.252656   156.797891      0.040351
```

(rest deleted)

We now execute the bootstrap. We extract the fixed X and then make a matrix to save the results in. We repeat the bootstrap process 1000 times:

```
> x <- model.matrix( ~ Area+Elevation+Nearest+Scruz+
  Adjacent,gala)[,-1]
> bcoef <- matrix(0,1000,6)
> for(i in 1:1000){
+ newy <- predict(g) + residuals(g)[sample(30,rep=T)]
+ brg <- ltsreg(x,newy,nsamp="best")
+ bcoef[i,] <- brg$coef
+ }
```

It is not convenient to use the `nsamp="exact"` since that would require 1000 times the time it takes to make the original estimate. That is about four hours on our computer. Being impatient, we compromised and used the second best option of `nsamp="best"`. This likely means that our bootstrap estimates of variability will be somewhat on the high side. This illustrates a common practical difficulty with the bootstrap — it can take a long time to compute. Fortunately, this problem recedes as processor speeds increase.

We can make a 95% confidence interval for this parameter by taking the empirical quantiles:

```
> quantile(bcoef[,2],c(0.025,0.975))
  2.5%   97.5%
1.4906 1.6173
```

Zero lies outside this interval so we are confident that there is an area effect. We can get a better picture of the distribution by looking at the density and marking the confidence interval:

```
> plot(density(bcoef[,2]),xlab="Coefficient of Area",main="")
> abline(v=quantile(bcoef[,2],c(0.025,0.975)))
```

See Figure 6.4. We see that the distribution is more peaked than a normal with some longish tails.

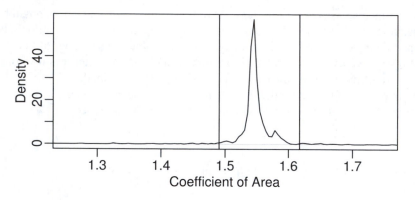

Figure 6.4 *Bootstrap distribution of* $\hat{\beta}_{Area}$ *with 95% confidence intervals.*

This would be more accurate if we took more than 1000 bootstrap resamples. The conclusion here would be that the area variable is significant. That is in contrast to the conclusion from the least squares fit. Which estimates are best? An examination of the Cook distances for the least squares fit shows the island of Isabela to be very influential. If we exclude this island from the least squares fit, we find that:

```
> gli <- lm(Species ~ Area + Elevation + Nearest + Scruz +
  Adjacent,  gala,subset=(row.names(gala) != "Isabela"))
> summary(gli)
Coefficients:
            Estimate Std. Error t value Pr(>|t|)
(Intercept)  22.5861    13.4019    1.69   0.1055
Area          0.2957     0.0619    4.78   8.0e-05
Elevation     0.1404     0.0497    2.82   0.0096
Nearest      -0.2552     0.7217   -0.35   0.7269
Scruz        -0.0901     0.1498   -0.60   0.5534
Adjacent     -0.0650     0.0122   -5.32   2.1e-05

Residual standard error: 41.6 on 23 degrees of freedom
Multiple R-Squared: 0.871,        Adjusted R-squared: 0.843
F-statistic: 31.2 on 5 and 23 DF,   p-value: 1.62e-09
```

This fit is much closer to the LTS fit in that Area and Adjacent are very significant predictors. Thus, there are two routes to the same goal. We can use regression diagnostics in conjunction with least squares to identify bad or unusual points or we can use robust methods. The former approach is more flexible and allows for the discovery of a wider class of problems, but it is time consuming and does require human intervention. When data need to be quickly analyzed, perhaps without expert assistance or when large numbers of datasets need to be fitted, robust methods give some protection against aberrant data.

Another interesting point is that the M-estimate failed to identify the unusual island, Isabela, and gave similar results to the full data least squares fit. We can show similar behavior on another dataset — consider the Star data presented in Section 4.2.2. We compute the least squares, Huber and LTS fits and display them in Figure 6.5:

```
> data(star)
> plot(light ~ temp, star)
> gs1 <- lm(light ~ temp, star)
> abline(coef(gs1))
> gs2 <- rlm(light ~ temp, star)
> abline(coef(gs2), lty=2)
> gs3 <- ltsreg(light ~ temp, star, nsamp="exact")
> abline(coef(gs3), lty=5)
```

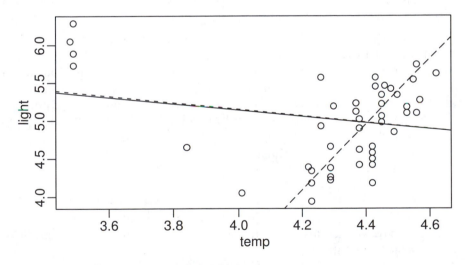

Figure 6.5 *Regression fits compared. Least squares is the solid line, Huber is the dotted line and LTS is the dashed line. Only LTS finds the fit in the bulk of the data.*

Only LTS managed to capture the trend in the main group of points. The Huber estimate is almost the same as the least squares estimate.

Summary

1. Robust estimators provide protection against long-tailed errors, but they cannot overcome problems with the choice of model and its variance structure.

2. Robust estimates just give you $\hat{\beta}$ and possibly standard errors without the associated inferential methods. Software and methodology for this inference is not easy to come by. The bootstrap is a general-purpose inferential method which is useful in these situations.

3. Robust methods can be used in addition to least squares as a confirmatory method. You have cause to worry if the two estimates are far apart. The source of the difference should be investigated.

4. Robust estimates are useful when data need to be fit automatically without the intervention of a skilled analyst.

Exercises

1. Researchers at National Institutes of Standards and Technology (NIST) collected pipeline data on ultrasonic measurements of the depths of defects in the Alaska pipeline in the field. The depth of the defects were then remeasured in the laboratory. These measurements were performed in six different batches. It turns out that this batch effect is not significant and so can be ignored in the analysis that follows. The laboratory measurements are more accurate than the in-field measurements, but more time consuming and expensive. We want to develop an regression equation for correcting the in-field measurements.

   (a) Fit a regression model Lab ˜ Field. Check for nonconstant variance.

   (b) We wish to use weights to account for the nonconstant variance. Here we split the range of Field into 12 groups of size nine (except for the last group which has only eight values). Within each group, we compute the variance of Lab as varlab and the mean of Field as meanfield. Supposing pipeline is the name of your data frame, the following R code will make the needed computations:

   ```
   > i <- order(pipeline$Field)
   > npipe <- pipeline[i,]
   > ff <- gl(12,9)[-108]
   > meanfield <- unlist(lapply(split(npipe$Field,ff),mean))
   > varlab <- unlist(lapply(split(npipe$Lab,ff),var))
   ```

   Suppose we guess that the the variance in the response is linked to the predictor in the following way:

   $$var(Lab) = a_0 Field^{a_1}$$

   Regress log(varlab) on log(meanfield) to estimate a_0 and a_1. (You might choose to remove the last point.) Use this to determine appropriate weights in a WLS fit of Lab on Field. Show the regression summary.

   (c) An alternative to weighting is transformation. Find transformations on Lab and/or Field so that in the transformed scale the relationship is approximately linear with constant variance. You may restrict your choice of transformation to square root, log and inverse.

2. Using the `divusa` data, fit a regression model with `divorce` as the response and `unemployed, femlab, marriage, birth` and `military` as predictors.

(a) Make two graphical checks for correlated errors. What do you conclude?

(b) Allow for serial correlation with an AR(1) model for the errors. (Hint: Use maximum likelihood to estimate the parameters in the GLS fit by `gls(...,method="ML", ...)`. What is the estimated correlation and is it significant? Does the GLS model change which variables are found to be significant?

(c) Speculate why there might be correlation in the errors.

3. For the `salmonella` dataset, fit a linear model with `colonies` as the response and `log(dose+1)` as the predictor. Check for lack of fit.

4. For the `cars` dataset, fit a linear model with `distance` as the response and `speed` as the predictor. Check for lack of fit.

5. Using the `stackloss` data, fit a model with `stack.loss` as the response and the other three variables as predictors using the following methods:

(a) Least squares

(b) Least absolute deviations

(c) Huber method

(d) Least trimmed squares

Compare the results. Now use diagnostic methods to detect any outliers or influential points. Remove these points and then use least squares. Compare the results.

CHAPTER 7

Transformation

Transformations of the response and predictors can improve the fit and correct violations of model assumptions such as nonconstant error variance. We may also consider adding additional predictors that are functions of the existing predictors like quadratic or cross-product terms. This means we have more choice in choosing the transformations on the predictors than on the response.

7.1 Transforming the Response

We start with some general considerations about transforming the response. Suppose that you are contemplating a logged response in a simple regression situation:

$$\log y = \beta_0 + \beta_1 x + \varepsilon$$

In the original scale of the response, this model becomes:

$$y = \exp(\beta_0 + \beta_1 x) \cdot \exp(\varepsilon)$$

In this model, the errors enter *multiplicatively* and not *additively* as they usually do. So the use of standard regression methods for the logged response model requires that we believe the errors enter multiplicatively in the original scale. Notice that if we believe the true model for y to be:

$$y = \exp(\beta_0 + \beta_1 x) + \varepsilon$$

then we cannot linearize this model and nonlinear regression methods would need to be applied.

As a practical matter, we usually do not know how the errors enter the model, additively, multiplicatively or otherwise. The typical approach is to try different transforms and then check the residuals to see whether they satisfy the conditions required for linear regression. Unless you have good information that the error enters in some particular way, this is the simplest and most appropriate approach.

Although you may transform the response, you will probably need to express predictions in the original scale. This is simply a matter of back transforming. For example, in the logged model above, your prediction would be $\exp(\hat{y}_0)$. If your prediction confidence interval in the logged scale was $[l, u]$, then you would use $[\exp l, \exp u]$. This interval will not be symmetric, but this may be desirable. For example, the untransformed prediction intervals for the Galápagos data went below zero in Section 3.5. Transformation of the response avoids this problem.

Regression coefficients will need to be interpreted with respect to the transformed scale. There is no straightforward way of back transforming them to values that can be interpreted in the original scale. You cannot directly compare regression coefficients for models where the response transformation is different. Difficulties of this type may dissuade one from transforming the response even if this requires the use of another type of model, such as a generalized linear model.

When you use a log transformation on the response, the regression coefficients have a particular interpretation:

$$\log \hat{y} = \hat{\beta}_0 + \hat{\beta}_1 x_1 + \cdots + \hat{\beta}_p x_p$$
$$\hat{y} = e^{\hat{\beta}_0} e^{\hat{\beta}_1 x_1} \cdots e^{\hat{\beta}_p x_p}$$

An increase of one in x_1 would multiply the predicted response (in the original scale) by $e^{\hat{\beta}_1}$. Thus when a log scale is used, the regression coefficients can be interpreted in a multiplicative rather than an additive manner.

The Box–Cox method is a popular way to determine a transformation on the response. It is designed for strictly positive responses and chooses the transformation to find the best fit to the data. The method transforms the response $y \rightarrow g_\lambda(y)$ where the family of transformations indexed by λ is:

$$g_\lambda(y) = \begin{cases} \frac{y^\lambda - 1}{\lambda} & \lambda \neq 0 \\ \log y & \lambda = 0 \end{cases}$$

For fixed $y > 0$, $g_\lambda(y)$ is continuous in λ. Choose λ using maximum likelihood. The profile log-likelihood assuming normality of the errors is:

$$L(\lambda) = -\frac{n}{2} \log(\text{RSS}_\lambda / n) + (\lambda - 1) \sum \log y_i$$

where RSS_λ is the residual sum of squares when $g_\lambda(y)$ is the response. You can compute $\hat{\lambda}$ numerically to maximize this. If the purpose of the regression model is just prediction, then use y^λ as the response (no need to use $(y^\lambda - 1)/\lambda$, as the rescaling is just for convenience in maximizing the likelihood). If explaining the model is important, you should round λ to the nearest interpretable value. For example, if $\hat{\lambda} = 0.46$, it would be hard to explain what this new response means, but $\sqrt{y}$ might be easier.

Transforming the response can make the model harder to interpret so we do not want to do it unless it is really necessary. One way to check this is to form a confidence interval for λ. A $100(1 - \alpha)\%$ confidence interval for λ is:

$$\{\lambda : \quad L(\lambda) > L(\hat{\lambda}) - \frac{1}{2}\chi_1^{2^{(1-\alpha)}}\}$$

This interval can be derived by inverting the likelihood ratio test of the hypothesis that $H_0 : \lambda = \lambda_0$ which uses the statistic $2(L(\hat{\lambda}) - L(\lambda_0))$ having approximate null distribution χ_1^2. The confidence interval also tells you how much it is reasonable to round λ for the sake of interpretability.

We check whether the response in the savings data need transformations. We will need the `boxcox` function from the MASS package:

```
> library(MASS)
```

Try it out on the savings dataset and plot the results:

```
> data(savings)
> g <- lm(sr ~ pop15+pop75+dpi+ddpi,savings)
> boxcox(g,plotit=T)
> boxcox(g,plotit=T,lambda=seq(0.5,1.5,by=0.1))
```

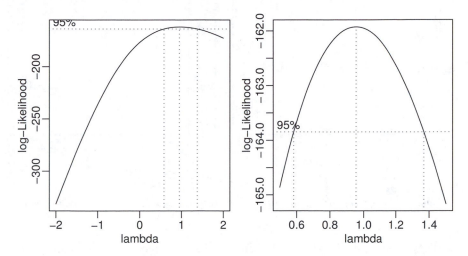

Figure 7.1 *Log-likelihood plots for the Box–Cox transformation of the savings data.*

The first plot shown in Figure 7.1 is too broad. I narrowed the range of λ in the second plot so that we can read off the confidence interval more easily.

The confidence interval for λ runs from about 0.6 to about 1.4. We can see that there is no good reason to transform.

Now consider the Galápagos Islands dataset analyzed earlier:

```
> data(gala)
> g <- lm(Species ~ Area + Elevation + Nearest + Scruz +
  Adjacent,gala)
> boxcox(g,plotit=T)
> boxcox(g,lambda=seq(0.0,1.0,by=0.05),plotit=T)
```

The plots are shown in Figure 7.2. We see that perhaps a cube root transformation might be best here. A square root is also a possibility, as this falls just within the confidence intervals. Certainly there is a strong need to transform.

Some general considerations concerning the Box–Cox method are:

1. The Box–Cox method gets upset by outliers — if you find $\hat{\lambda} = 5$, then this is probably the reason — there can be little justification for actually making such an extreme transformation.

2. If some $y_i < 0$, we can add a constant to all the y. This can work provided the constant is small, but this is an inelegant solution.

3. If $\max_i y_i / \min_i y_i$ is small, then the Box–Cox will not have much real effect because power transforms are well approximated by linear transformations over short intervals far from the origin.

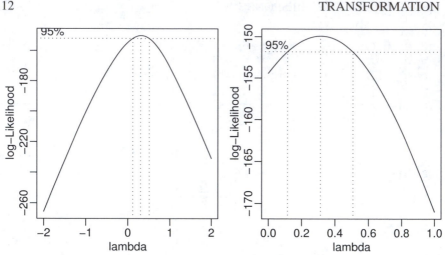

Figure 7.2 *Log-likelihood plots for the Box–Cox transformation of the Galápagos data.*

4. There is some doubt whether the estimation of λ counts as an extra parameter to be considered in the degrees of freedom. This is a difficult question since λ is not a linear parameter and its estimation is not part of the least squares fit.

The Box–Cox method is not the only way of transforming the predictors. For responses that are proportions (or percentages), the logit transformation, $\log(y/(1 - y))$, is often used, while for responses that are correlations, Fisher's z transform, $y = 0.5\log((1+y)/(1-y))$, is worth considering.

7.2 Transforming the Predictors

You can take a Box–Cox style approach for each of the predictors, choosing the transformation to minimize the RSS. However, this takes time. You can also use graphical methods such as partial residual plots to select transforming the predictors. These methods are designed to replace x in the model with f(x) for some chosen f. The methods we consider below are more general in that they replace x with more than one term — f(x) + g(x) + ... This allows more flexibility.

7.2.1 Broken Stick Regression

Sometimes we have reason to believe that different linear regression models apply in different regions of the data. For example, in the analysis of the savings data, we observed that there were two groups in the data and we might want to fit a different model to the two parts. Suppose we focus attention on just the pop15 predictor for ease of presentation. We fit the two regression models depending on whether pop15 is greater or less than 35%. The two fits are seen in Figure 7.3:

```
> g1 <- lm(sr ~ pop15, savings, subset=(pop15 < 35))
```

```
> g2 <- lm(sr ~ pop15, savings, subset=(pop15 > 35))
> plot(sr ~ pop15, savings, xlab="Pop'n under 15",
  ylab="Savings Rate")
> abline(v=35, lty=5)
> segments(20, g1$coef[1]+g1$coef[2]*20, 35,
  g1$coef[1]+g1$coef[2]*35)
> segments(48, g2$coef[1]+g2$coef[2]*48, 35,
  g2$coef[1]+g2$coef[2]*35)
```

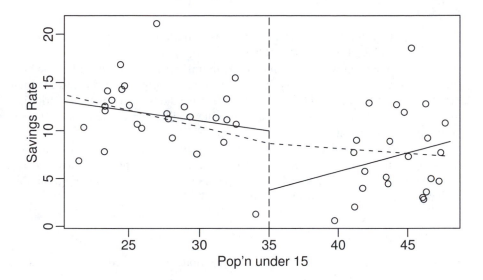

Figure 7.3 *Subset regression fit is shown with the solid line, while the broken stick regression is shown with the dotted line.*

A possible objection to this subsetted regression fit is that the two parts of the fit do not meet at the join. If we believe the fit should be continuous as the predictor varies, we should consider the broken stick regression fit. Define two *basis functions*:

$$B_l(x) = \begin{cases} c - x & \text{if } x < c \\ 0 & \text{otherwise} \end{cases}$$

and:

$$B_r(x) = \begin{cases} x - c & \text{if } x > c \\ 0 & \text{otherwise} \end{cases}$$

where c marks the division between the two groups. B_l and B_r form a first-order spline basis with a knotpoint at c. Sometimes B_l and B_r are called hockey-stick functions because of their shape. We can now fit a model of the form:

$$y = \beta_0 + \beta_1 B_l(x) + \beta_2 B_r(x) + \varepsilon$$

using standard regression methods. The two linear parts are guaranteed to meet at c. Notice that this model uses only three parameters in contrast to the four total parameters used in the subsetted regression illustrated before. A parameter has been saved by insisting on the continuity of the fit at *c*.

We define the two hockey-stick functions, compute and display the fit:

```
> lhs <- function(x) ifelse(x < 35,35-x,0)
> rhs <- function(x) ifelse(x < 35,0,x-35)
> gb <- lm(sr ~ lhs(pop15) + rhs(pop15), savings)
> x <- seq(20,48,by=1)
> py <- gb$coef[1]+gb$coef[2]*lhs(x)+gb$coef[3]*rhs(x)
> lines(x,py,lty=2)
```

The two (dotted) lines now meet at 35, as shown in Figure 7.3. The intercept of this model is the value of the response at the join.

We might question which fit is preferable in this particular instance. For the high pop15 countries, we see that the imposition of continuity causes a change in sign for the slope of the fit. We might argue that because the two groups of countries are so different and there are so few countries in the middle region, we might not want to impose continuity at all.

We can have more than one knotpoint simply by defining more basis functions with different knotpoints. Broken stick regression is sometimes called *segmented regression*. Allowing the knotpoints to be parameters is worth considering, but this will result in a nonlinear model.

7.2.2 Polynomials

Another way of generalizing the $X\beta$ part of the model is to add polynomial terms. In the one-predictor case, we have:

$$y = \beta_0 + \beta_1 x + \cdots + \beta_d x^d + \varepsilon$$

which allows for a more flexible relationship, although we usually do not believe it exactly represents any underlying reality.

There are two ways to choose d. We can keep adding terms until the added term is not statistically significant. Alternatively, we can start with a large d and eliminate nonstatistically significant terms starting with the highest order term.

Do *not* eliminate lower order terms from the model even if they are not statistically significant. An additive change in scale would change the t-statistic of all but the highest order term. We would not want the conclusions of our study to be so brittle to such changes in the scale which ought to be inconsequential.

Let's see if we can use polynomial regression on the ddpi variable in the savings data. First, fit a linear model:

```
> summary(lm(sr ~ ddpi,savings))
Coefficients:
            Estimate Std. Error t value Pr(>|t|)
(Intercept)    7.883     1.011     7.80   4.5e-10
ddpi           0.476     0.215     2.22    0.031
```

The p-value of ddpi is significant so move on to a quadratic term:

```
> summary(lm(sr ~ ddpi+I(ddpi^2),savings))
Coefficients:
            Estimate Std. Error t value Pr(>|t|)
(Intercept)   5.1304     1.4347    3.58  0.00082
ddpi          1.7575     0.5377    3.27  0.00203
I(ddpi^2)    -0.0930     0.0361   -2.57  0.01326
```

Again the p-value of $ddpi^2$ is significant so move on to a cubic term:

```
> summary(lm(sr ~ ddpi+I(ddpi^2)+I(ddpi^3),savings))
Coefficients:
            Estimate  Std. Error t value Pr(>|t|)
(Intercept)  5.145360   2.198606    2.34    0.024
ddpi         1.746017   1.380455    1.26    0.212
I(ddpi^2)   -0.090967   0.225598   -0.40    0.689
I(ddpi^3)   -0.000085   0.009374   -0.01    0.993
```

The p-value of $ddpi^3$ is not significant so stick with the quadratic. Notice how the other p-values are not significant in contrast with earlier results. Note that starting from a large model (including the fourth power) and working downwards gives the same result.

To illustrate the point about the significance of lower order terms, suppose we transform ddpi by subtracting 10 and refit the quadratic model:

```
> savings <- data.frame(savings,mddpi=savings$ddpi-10)
> summary(lm(sr ~ mddpi+I(mddpi^2),savings))
Coefficients:
            Estimate Std. Error t value Pr(>|t|)
(Intercept) 13.4070     1.4240    9.41  2.2e-12
mddpi       -0.1022     0.3027   -0.34    0.737
I(mddpi^2)  -0.0930     0.0361   -2.57    0.013
```

We see that the quadratic term remains unchanged, but the linear term is now insignificant. Since there is often no necessary importance to zero on a scale of measurement, there is no good reason to remove the linear term in this model but not in the previous version. No advantage would be gained.

You have to refit the model each time a term is removed and for large d there can be a problem with numerical stability. Orthogonal polynomials get around this problem by defining:

$$
\begin{aligned}
z_1 &= a_1 + b_1 x \\
z_2 &= a_2 + b_2 x + c_2 x^2 \\
z_3 &= a_3 + b_3 x + c_3 x^2 + d_3 x^3
\end{aligned}
$$

etc. where the coefficients $a, b, c, \ldots$ are chosen so that $z_i^T z_j = 0$ when $i \neq j$. The z are called orthogonal polynomials. The value of orthogonal polynomials has declined with advances in computing speeds although they are still worth knowing about because of their numerical stability and ease of use. The poly() function constructs orthogonal polynomials:

```
> g <- lm(sr ~ poly(ddpi,4),savings)
> summary(g)
```

```
Coefficients:
                Estimate Std. Error t value Pr(>|t|)
(Intercept)       9.6710     0.5846   16.54   <2e-16
poly(ddpi, 4)1    9.5590     4.1338    2.31    0.025
poly(ddpi, 4)2  -10.4999     4.1338   -2.54    0.015
poly(ddpi, 4)3   -0.0374     4.1338   -0.01    0.993
poly(ddpi, 4)4    3.6120     4.1338    0.87    0.387
```

```
Residual standard error: 4.13 on 45 degrees of freedom
Multiple R-Squared: 0.218,        Adjusted R-squared: 0.149
F-statistic: 3.14 on 4 and 45 DF,  p-value: 0.0232
```

We can come to the same conclusion as above, that the quadratic model is best, with just this summary.

You can also define polynomials in more than one variable. These are sometimes called *response surface* models. A second degree model would be:

$$y = \beta_0 + \beta_1 x_1 + \beta_2 x_2 + \beta_{11} x_1^2 + \beta_{22} x_2^2 + \beta_{12} x_1 x_2$$

an example of which could be fit as:

```
> g <- lm(sr ~ polym(pop15,ddpi,degree=2),savings)
```

7.2.3 Regression Splines

Polynomials have the advantage of smoothness, but the disadvantage that each data point affects the fit globally. This is because the power functions used for the polynomials take nonzero values across the whole range of the predictor. In contrast, the broken stick regression method localizes the influence of each data point to its particular segment which is good, but we do not have the same smoothness as with the polynomials. There is a way we can combine the beneficial aspects of both these methods — smoothness and local influence — by using *B-spline* basis functions.

We may define a cubic B-spline basis on the interval $[a,b]$ by the following requirements on the interior basis functions with knotpoints at $t_1, \ldots, t_k$:

1. A given basis function is nonzero on an interval defined by four successive knots and zero elsewhere. This property ensures the local influence property.

2. The basis function is a cubic polynomial for each subinterval between successive knots.

3. The basis function is continuous and is also continuous in its first and second derivatives at each knotpoint. This property ensures the smoothness of the fit.

4. The basis function integrates to one over its support.

The basis functions at the ends of the interval are defined a little differently to ensure continuity in derivatives at the edge of the interval. A full definition of B-splines and more details about their properties may be found in de Boor (2002). The broken stick regression is an example of the use of linear splines.

Let's see how the competing methods do on a constructed example. Suppose we know the true model is:

$$y = \sin^3(2\pi x^3) + \varepsilon, \qquad \varepsilon \sim N(0, (0.1)^2)$$

The advantage of using simulated data is that we can see how close our methods come to the truth. We generate the data and display them in the first plot of Figure 7.4.

```
> funky <- function(x) sin(2*pi*x^3)^3
> x <- seq(0,1,by=0.01)
> y <- funky(x) + 0.1*rnorm(101)
> matplot(x,cbind(y,funky(x)),type="pl",ylab="y",pch=18,lty=1)
```

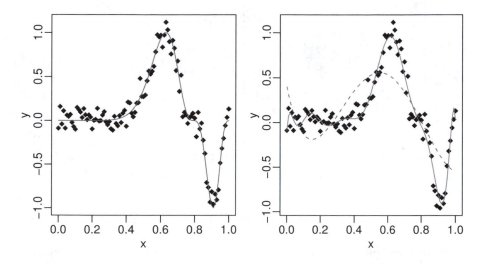

Figure 7.4 *Data and true function shown on the left. Orthogonal polynomial of order 4 (dashed) and order 12 (solid) shown on the right.*

We see how orthogonal polynomial bases of orders 4 and 12 do in fitting these data:

```
> g4 <- lm(y ~ poly(x,4))
> g12 <- lm(y ~ poly(x,12))
> matplot(x,cbind(y,g4$fit,g12$fit),type="pll",
  ylab="y",pch=18,lty=c(1,2))
```

The two fits are shown in the second panel of Figure 7.4. We see that order 4 is a clear underfit; order 12 is much better although the fit is too wiggly in the first section and misses the point of inflection around $x = 0.8$.

We now create the B-spline basis. You need to have three additional knots at the start and end to get the right basis. I have chosen to the knot locations to put more in the regions of greater curvature. I have used 12 basis functions for comparability to the orthogonal polynomial fit:

```
> library(splines)
> knots <- c(0,0,0,0,0.2,0.4,0.5,0.6,0.7,0.8,0.85,0.9,1,1,1,1)
> bx <- splineDesign(knots,x)
```

```
> gs <- lm(y ~ bx)
> matplot(x,bx,type="l")
> matplot(x,cbind(y,gs$fit),type="pl",ylab="y",pch=18,lty=1)
```

The basis functions themselves are seen in the first panel of Figure 7.5 while the fit
itself appears in the second panel. We see that the fit comes very close to the truth.

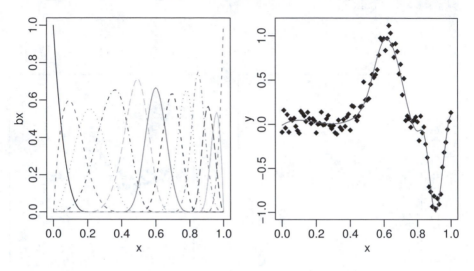

Figure 7.5 *Cubic spline basis function on the left, cubic spline fit to the data on the right.*

Regression splines are useful for fitting functions with some flexibility provided
we have enough data. We can form basis functions for all the predictors in our model
but we need to be careful not to use up too many degrees of freedom.

7.2.4 Overview

The methods described above are somewhat awkward to apply exhaustively and even
then they may miss important structure because of the problem of trying to find
good transformations on several variables simultaneously. One recent approach is
the additive model:

$$y = \beta_0 + f_1(x_1) + f_2(x_2) + \cdots + f_p(x_p) + \varepsilon$$

where nonparametric regression techniques are used to estimate the f_is. Alternatively, you could implement this using the regression spline bases for each predictor
variable. Other techniques are ACE, AVAS, regression trees, MARS and neural networks. See, for example, Ripley (1996) and Hastie, Tibshirani, and Friedman (2001)
among many others.

It is important to realize the strengths and weaknesses of regression analysis. For
larger datasets with relatively little noise, more recently developed complex models
will be able to fit the data better while keeping the number of parameters under
control. For smaller datasets or where the noise level is high (as is typically found in

the social sciences), more complex models are not justified and standard regression is most effective. One relative advantage of regression is that the models are easier to interpret in contrast to techniques like neural networks which are usually only good for predictive purposes.

Exercises

1. The `aatemp` data come from the U.S. Historical Climatology network. They are the annual mean temperatures (in degrees F) in Ann Arbor, Michigan going back about 150 years.

   (a) Is there a linear trend?

   (b) Observations in successive years may be correlated. Fit a model that estimates this correlation. Does this change your opinion about the trend?

   (c) Fit a polynomial model with degree 10 and use backward elimination to reduce the degree of the model. Plot your fitted model on top of the data. Use this model to predict the temperature in 2020.

   (d) Suppose someone claims that the temperature was constant until 1930 and then began a linear trend. Fit a model corresponding to this claim. What does the fitted model say about this claim?

   (e) Make a cubic spline fit with six basis functions evenly spaced on the range. Plot the fit in comparison to the previous fits. Does this model fit better than the straight-line model?

2. The `cornnit` data on the relationship between corn yield (bushels per acre) and nitrogen (pounds per acre) fertilizer application were studied in Wisconsin in 1994. Use transformations to find a good model for predicting yield from nitrogen. Use a goodness of fit test to check your model.

3. Using the `ozone` data, fit a model with O3 as the response and `temp, humidity` and `ibh` as predictors. Use the Box–Cox method to determine the best transformation on the response.

4. Using the `pressure` data, fit a model with `pressure` as the response and `temperature` as the predictor using transformations to obtain a good fit.

5. Use transformations to find a good model for `volume` in terms of `girth` and `height` using the `trees` data.

CHAPTER 8

Variable Selection

Variable selection is intended to select the "best" subset of predictors. Several reasons for wanting to do this follow:

1. We want to explain the data in the simplest way. Redundant predictors should be removed. The principle of Occam's Razor states that among several plausible explanations for a phenomenon, the simplest is best. Applied to regression analysis, this implies that the smallest model that fits the data is best.

2. Unnecessary predictors will add noise to the estimation of other quantities that interested us. Degrees of freedom will be wasted. More precise estimates and predictions might be achieved with a smaller model.

3. Collinearity is caused by having too many variables trying to do the same job. Removing excess predictors will aid interpretation.

4. If the model is to be used for prediction, we can save time and/or money by not measuring redundant predictors.

Variable selection is a process that should not be separated from the rest of the analysis. Other parts of the data analysis can have an impact. For example, outliers and influential points can do more than just change the current model — they can change the model we select. It is important to identify such points. Also transformations of the variables can have an impact on the model selected. Some iteration and experimentation is often necessary to find better models.

There are two main types of variable selection. The stepwise testing approach compares successive models while the criterion approach attempts to find the model that optimizes some measure of goodness.

8.1 Hierarchical Models

Some models have a natural hierarchy. For example, in polynomial models, x^2 is a higher order term than x. When selecting variables, it is important to respect the hierarchy. Lower order terms should not be removed from the model before higher order terms in the same variable. There are two common situations where this can arise:

Consider the polynomial model:

$$y = \beta_0 + \beta_1 x + \beta_2 x^2 + \varepsilon$$

Suppose we fit this model and find that the regression summary shows that the term in x is not significant but the term in x^2 is. If we then remove the x term, our reduced model would become:

$$y = \beta_0 + \beta_2 x^2 + \varepsilon$$

However, suppose we make a scale change $x \rightarrow x + a$; then the model would become:

$$y = \beta_0 + \beta_2 a^2 + 2\beta_2 ax + \beta_2 x^2 + \varepsilon$$

The first order x term has now reappeared. Scale changes should not make any important change to the model, but in this case an additional term has been added. This is not desirable. This illustrates why we should not remove lower order terms in the presence of higher order terms. We would not want interpretation to depend on the choice of scale. Removal of the first-order term here corresponds to the hypothesis that the predicted response is symmetric about and has an optimum at $x = 0$. Usually this hypothesis is not meaningful and should not be considered. Only when this hypothesis makes sense in the context of the particular problem could we justify the removal of the lower order term.

For models with interactions, consider the example of a second order response surface model:

$$y = \beta_0 + \beta_1 x_1 + \beta_2 x_2 + \beta_{11} x_1^2 + \beta_{22} x_2^2 + \beta_{12} x_1 x_2$$

We would not normally consider removing the $x_1 x_2$ interaction term without simultaneously considering the removal of the x_1^2 and x_2^2 terms. A joint removal would correspond to the clearly meaningful comparison of a quadratic surface and a linear one. Just removing the $x_1 x_2$ term would correspond to a surface that is aligned with the coordinate axes. This is hard to interpret and should not be considered unless some particular meaning can be attached. Any rotation of the predictor space would reintroduce the interaction term and, as with the polynomials, we would not ordinarily want our model interpretation to depend on the particular basis for the predictors.

8.2 Testing-Based Procedures

8.2.1 Backward Elimination

This is the simplest of all variable selection procedures and can be easily implemented without special software. In situations where there is a complex hierarchy, backward elimination can be run manually while taking account of what variables are eligible for removal.

We start with all the predictors in the model and then remove the predictor with highest p-value greater than α_{crit}. Next refit the model and remove the remaining least significant predictor provided its p-value is greater than α_{crit}. Sooner or later, all "nonsignificant" predictors will be removed and the selection process will be complete.

The α_{crit} is sometimes called the "p-to-remove" and does not have to be 5%. If prediction performance is the goal, then a 15 to 20% cutoff may work best, although methods designed more directly for optimal prediction should be preferred.

8.2.2 Forward Selection

This just reverses the backward method. We start with no variables in the model and then for all predictors not in the model, we check their p-value if they are added to the model. We choose the one with lowest p-value less than α_{crit}. We continue until no new predictors can be added.

8.2.3 Stepwise Regression

This is a combination of backward elimination and forward selection. This addresses the situation where variables are added or removed early in the process and we want to change our mind about them later. At each stage a variable may be added or removed and there are several variations on exactly how this is done.

8.2.4 Verdict on Testing-Based Methods

Testing-based procedures are relatively cheap computationally, but they do have some of the following drawbacks:

1. Because of the "one-at-a-time" nature of adding/dropping variables, it is possible to miss the "optimal" model.

2. The p-values used should not be treated too literally. There is so much multiple testing occurring that the validity is dubious. The removal of less significant predictors tends to increase the significance of the remaining predictors. This effect leads one to overstate the importance of the remaining predictors.

3. The procedures are not directly linked to final objectives of prediction or explanation and so may not really help solve the problem of interest. With any variable selection method, it is important to keep in mind that model selection cannot be divorced from the underlying purpose of the investigation. Variable selection tends to amplify the statistical significance of the variables that stay in the model. Variables that are dropped can still be correlated with the response. It would be wrong to say that these variables are unrelated to the response; it is just that they provide no additional explanatory effect beyond those variables already included in the model.

4. Stepwise variable selection tends to pick models that are smaller than desirable for prediction purposes. To give a simple example, consider the simple regression with just one predictor variable. Suppose that the slope for this predictor is not quite statistically significant. We might not have enough evidence to say that it is related to y but it still might be better to use it for predictive purposes.

We illustrate the variable selection methods on some data on the 50 states from the 1970s. The data were collected from U.S. Bureau of the Census. We will take life expectancy as the response and the remaining variables as predictors:

```
> data(state)
> statedata <- data.frame(state.x77,row.names=state.abb)
> g <- lm(Life.Exp ~ ., data=statedata)
```

```
> summary(g)
Coefficients:
               Estimate Std. Error t value Pr(>|t|)
(Intercept)    7.09e+01   1.75e+00   40.59  < 2e-16
Population     5.18e-05   2.92e-05    1.77    0.083
Income        -2.18e-05   2.44e-04   -0.09    0.929
Illiteracy     3.38e-02   3.66e-01    0.09    0.927
Murder        -3.01e-01   4.66e-02   -6.46  8.7e-08
HS.Grad        4.89e-02   2.33e-02    2.10    0.042
Frost         -5.74e-03   3.14e-03   -1.82    0.075
Area          -7.38e-08   1.67e-06   -0.04    0.965

Residual standard error: 0.745 on 42 degrees of freedom
Multiple R-Squared: 0.736,        Adjusted R-squared: 0.692
F-statistic: 16.7 on 7 and 42 DF,  p-value: 2.53e-10
```

The signs of some of the coefficients match plausible expectations concerning how the predictors might affect the response. Higher murders rate decrease life expectancy as one might expect. Even so, some variables such as income, are not significant, contrary to what one might expect.

We illustrate the backward method — at each stage we remove the predictor with the largest p-value over 0.05:

```
> g <- update(g, . ~ . - Area)
> summary(g)
Coefficients:
               Estimate Std. Error t value Pr(>|t|)
(Intercept)    7.10e+01   1.39e+00   51.17  < 2e-16
Population     5.19e-05   2.88e-05    1.80    0.079
Income        -2.44e-05   2.34e-04   -0.10    0.917
Illiteracy     2.85e-02   3.42e-01    0.08    0.934
Murder        -3.02e-01   4.33e-02   -6.96  1.5e-08
HS.Grad        4.85e-02   2.07e-02    2.35    0.024
Frost         -5.78e-03   2.97e-03   -1.94    0.058

> g <- update(g, . ~ . - Illiteracy)
> summary(g)

Coefficients:
               Estimate Std. Error t value Pr(>|t|)
(Intercept)    7.11e+01   1.03e+00   69.07  < 2e-16
Population     5.11e-05   2.71e-05    1.89    0.066
Income        -2.48e-05   2.32e-04   -0.11    0.915
Murder        -3.00e-01   3.70e-02   -8.10  2.9e-10
HS.Grad        4.78e-02   1.86e-02    2.57    0.014
Frost         -5.91e-03   2.47e-03   -2.39    0.021

> g <- update(g, . ~ . - Income)
> summary(g)
```

```
Coefficients:
             Estimate Std. Error t value Pr(>|t|)
(Intercept)  7.10e+01    9.53e-01   74.54   < 2e-16
Population   5.01e-05    2.51e-05    2.00    0.0520
Murder      -3.00e-01    3.66e-02   -8.20   1.8e-10
HS.Grad      4.66e-02    1.48e-02    3.14    0.0030
Frost       -5.94e-03    2.42e-03   -2.46    0.0180

> g <- update(g, . ˜ . - Population)
> summary(g)

Coefficients:
             Estimate Std. Error t value Pr(>|t|)
(Intercept) 71.03638    0.98326   72.25    <2e-16
Murder      -0.28307    0.03673   -7.71    8e-10
HS.Grad      0.04995    0.01520    3.29    0.0020
Frost       -0.00691    0.00245   -2.82    0.0070

Residual standard error: 0.743 on 46 degrees of freedom
Multiple R-Squared: 0.713,       Adjusted R-squared: 0.694
F-statistic:   38 on 3 and 46 DF,  p-value: 1.63e-12
```

The final removal of the Population variable is a close call. We may want to consider including this variable if interpretation is made easier. Notice that the R^2 for the full model of 0.736 is reduced only slightly to 0.713 in the final model. Thus the removal of four predictors causes only a minor reduction in fit.

It is important to understand that the variables omitted from the model may still be related to the response. For example:

```
> summary(lm(Life.Exp ˜ Illiteracy+Murder+Frost, statedata))
Coefficients:
             Estimate Std. Error t value Pr(>|t|)
(Intercept) 74.55672    0.58425  127.61    <2e-16
Illiteracy  -0.60176    0.29893   -2.01    0.0500
Murder      -0.28005    0.04339   -6.45    6e-08
Frost       -0.00869    0.00296   -2.94    0.0052

Residual standard error: 0.791 on 46 degrees of freedom
Multiple R-Squared: 0.674,       Adjusted R-squared: 0.653
F-statistic: 31.7 on 3 and 46 DF,  p-value: 2.91e-11
```

we see that Illiteracy does have some association with Life Expectancy. It is true that replacing Illiteracy with High School graduation rate gives us a somewhat better fitting model, but it would be insufficient to conclude that Illiteracy is not a variable of interest.

8.3 Criterion-Based Procedures

If we have some idea about the purpose for which a model is intended, we might propose some measure of how well a given model meets that purpose. We could choose that model among those possible that optimizes that criterion. If there are q potential predictors, then there are 2^q possible models. We could fit all these models and choose the best one according to some criterion. For larger q, this might be too time consuming and we may need to economize by limiting the search. Some possible criteria are the Akaike Information Criterion (AIC), defined as -2 max log-likelihood $+ 2p$. Also used is the Bayes Information Criterion (BIC) which is -2 max log-likelihood $+ p \log n$. For linear regression models, the -2 max loglikelihood is $n \log(RSS/n)+$ a constant. Since the constant is the same for a given data set and assumed error distribution, it can be ignored for regression model comparisons on the same data. Additional care is necessary for other types of comparisons.

We want to minimize AIC or BIC. Larger models will fit better and so have smaller residual sum of squares (RSS), but use more parameters. Thus the best model choice will balance fit with model size. BIC penalizes larger models more heavily and so will tend to prefer smaller models in comparison to AIC. AIC and BIC are often used as selection criteria for other types of models too.

We can apply the AIC (and optionally the BIC) to the state data. The function does not evaluate the AIC for all possible models but uses a search method that compares models sequentially. Thus it bears some comparison to the stepwise method described above, but only in the method of search — there is no hypothesis testing.

```
> g <- lm(Life.Exp ~ ., data=statedata)
> step(g)
Start:  AIC= -22.18
 Life.Exp ~ Population + Income + Illiteracy + Murder +
    HS.Grad + Frost + Area

              Df Sum of Sq    RSS    AIC
- Area         1     0.0011  23.3  -24.2
- Income       1     0.0044  23.3  -24.2
- Illiteracy   1     0.0047  23.3  -24.2
<none>                       23.3  -22.2
- Population   1        1.7  25.0  -20.6
- Frost        1        1.8  25.1  -20.4
- HS.Grad      1        2.4  25.7  -19.2
- Murder       1       23.1  46.4   10.3

Step:  AIC= -24.18
 Life.Exp ~ Population + Income + Illiteracy + Murder +
    HS.Grad + Frost

.. intermediate steps omitted ..

Step:  AIC= -28.16
 Life.Exp ~ Population + Murder + HS.Grad + Frost

              Df Sum of Sq    RSS    AIC
<none>                       23.3  -28.2
- Population   1        2.1  25.4  -25.9
```

```
- Frost          1       3.1  26.4 -23.9
- HS.Grad        1       5.1  28.4 -20.2
- Murder         1      34.8  58.1  15.5


Coefficients:
(Intercept)    Population      Murder    HS.Grad      Frost
   7.10e+01      5.01e-05   -3.00e-01   4.66e-02  -5.94e-03
```

The sequence of variable removal is the same as with backward elimination. The only difference is that the population variable is retained.

Another commonly used criterion is adjusted R^2, written R_a^2. Recall that $R^2 = 1 - RSS/TSS$. Adding a variable to a model can only decrease the RSS and so only increase the R^2. Hence R^2 by itself is not a good criterion, because it would always choose the largest possible model:

$$R_a^2 = 1 - \frac{RSS/(n-p)}{TSS/(n-1)} = 1 - \left(\frac{n-1}{n-p}\right)(1-R^2) = 1 - \frac{\hat{\sigma}_{model}^2}{\hat{\sigma}_{null}^2}$$

Adding a predictor will only increase R_a^2 if it has some predictive value. There is a connection to $\hat{\sigma}^2$. Minimizing the standard error for prediction means minimizing $\hat{\sigma}^2$ which in term means maximizing R_a^2.

Our final criterion is Mallow's C_p statistic. A good model should predict well, so the average mean square error of prediction might be a good criterion:

$$\frac{1}{\sigma^2}\sum_i E(\hat{y}_i - Ey_i)^2$$

which can be estimated by the C_p statistic:

$$C_p = \frac{RSS_p}{\hat{\sigma}^2} + 2p - n$$

where $\hat{\sigma}^2$ is from the model with all predictors and RSS_p indicates the RSS from a model with p parameters. For the full model $C_p = p$ exactly. If a p predictor model fits, then $E(RSS_p) = (n-p)\sigma^2$ and then $E(C_p) \approx p$. A model with a bad fit will have C_p much bigger than p. It is usual to plot C_p against p. We desire models with small p and C_p around or less than p.

C_p, R_a^2 and AIC all trade-off fit in terms of RSS against complexity (p).

Now we try the C_p and R_a^2 methods for the selection of variables in the state dataset. For a model of a given size, all the methods above (with the possible exception of PRESS) will select the model with the smallest residual sum of squares. We first discover the best models for each size:

```
> library(leaps)
> b<-regsubsets(Life.Exp~.,data=statedata)
> (rs <- summary(b))
1 subsets of each size up to 7
Selection Algorithm: exhaustive
          Population Income Illiteracy Murder HS.Grad Frost Area
1  ( 1 )  " "        " "    " "        "*"    " "     " "   " "
2  ( 1 )  " "        " "    " "        "*"    "*"     " "   " "
```

3	(1)	" "		" "		" "		" * "	" * "	" * "	" "	" "
4	(1)	" * "		" "		" "		" * "	" * "	" * "	" "	" "
5	(1)	" * "		" * "		" "		" * "	" * "	" * "	" "	" "
6	(1)	" * "		" * "		" * "		" * "	" * "	" * "	" "	" "
7	(1)	" * "		" * "		" * "		" * "	" * "	" * "	" * "

In some cases, we might consider more than one model per size. Here we see that the best one predictor model uses `Murder` and so on. The C_p plot can be constructed as:

```
> plot(2:8,rs$cp,xlab="No. of Parameters",ylab="Cp Statistic")
> abline(0,1)
```

as seen in the first panel of Figure 8.1.

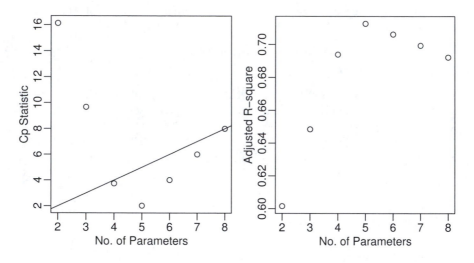

Figure 8.1 *The C_p plot for the state data on the left; adjusted R^2 on the right.*

The competition is between the four-parameter, three-predictor, model including Frost, HS graduation and Murder and the model also including Population. Both models are on or below the $C_p = p$ line, indicating good fits. The choice is between the smaller model and the larger model, which fits a little better. Some even larger models fit in the sense that they are on or below the $C_p = p$ line, but we would not opt for these in the presence of smaller models that fit.

Now let's see which model the adjusted R^2 criterion selects using the plot shown in the second panel of Figure 8.1:

```
> plot(2:8,rs$adjr2,xlab="No. of Parameters",
  ylab="Adjusted R-square")
```

We see that the Population, Frost, HS graduation and Murder model has the largest R_a^2.

Variable selection methods are sensitive to outliers and influential points. Let's check for high leverage points:

```
> h <- lm.influence(g)$hat
> names(h) <- state.abb
> rev(sort(h))
      AK        CA        HI        NV        NM        TX        NY
0.809522  0.408857  0.378762  0.365246  0.324722  0.284164  0.256950
```

We can see that Alaska has high leverage. Let's try excluding it:

```
> b<-regsubsets(Life.Exp~.,data=statedata,
  subset=(state.abb!="AK"))
> rs <- summary(b)
> rs$which[which.max(rs$adjr),]
(Intercept)  Population       Income  Illiteracy       Murder
       TRUE        TRUE        FALSE       FALSE         TRUE
   HS.Grad        Frost         Area
      TRUE         TRUE         TRUE
```

We see that `Area` now makes it into the model. Transforming the predictors can also have an effect. Take a look at the variables:

```
> stripchart(data.frame(scale(statedata)),vertical=TRUE,
  method="jitter")
```

Jittering adds a small amount of noise (in the horizontal direction in this example). It is useful for moving apart points that would otherwise overprint each other.

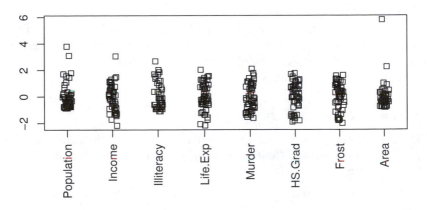

Figure 8.2 *Strip charts of the state data; all variables have been standardized.*

In Figure 8.2, we see that Population and Area are skewed — we try transforming them:

```
> b<-regsubsets(Life.Exp~log(Population)+Income+Illiteracy+
  Murder+HS.Grad+Frost+log(Area),statedata)
> rs <- summary(b)
> rs$which[which.max(rs$adjr),]
   (Intercept)  log(Population)              Income    Illiteracy
          TRUE             TRUE               FALSE         FALSE
```

Murder	HS.Grad	Frost	log(Area)
TRUE	TRUE	TRUE	FALSE

This changes the "best" model again to log(Population), Frost, HS graduation and Murder. The adjusted R^2 of 71.7% is the highest among models we have seen so far.

8.4 Summary

Variable selection is a means to an end and not an end itself. The aim is to construct a model that predicts well or explains the relationships in the data. Automatic variable selections are not guaranteed to be consistent with these goals. Use these methods as a guide only.

Stepwise methods use a restricted search through the space of potential models and use a dubious hypothesis testing-based method for choosing between models. Criterion-based methods typically involve a wider search and compare models in a preferable manner. For this reason, I recommend that you use a criterion-based method.

Accept the possibility that several models may be suggested which fit about as well as each other. If this happens, consider:

1. Do the models have similar qualitative consequences?
2. Do they make similar predictions?
3. What is the cost of measuring the predictors?
4. Which has the best diagnostics?

If you find models that seem roughly equally as good, but lead to quite different conclusions, then it is clear that the data cannot answer the question of interest unambiguously. Be alert to the possibility that a model contradictory to the tentative conclusions might be out there.

Exercises

1. Use the `prostate` data with `lpsa` as the response and the other variables as predictors. Implement the following variable selection methods to determine the "best" model:

   (a) Backward Elimination
   (b) AIC
   (c) Adjusted R^2
   (d) Mallows C_p

2. Using the `teengamb` dataset with `gamble` as the response and the other variables as predictors, repeat the work of the first question.

3. Using the `divusa` dataset with `divorce` as the response and the other variables as predictors, repeat the work of the first question.

4. Using the `trees` data, fit a model with `log(Volume)` as the response and a second-order polynomial (including the interaction term) in `Girth` and `Height`. Determine whether the model may be reasonably simplified.

5. Fit a linear model to the `stackloss` data with `stack.loss` as the predictor and the other variables as predictors. Simplify the model if possible. Check the model for outliers and influential points. Now return to the full model, determine whether there are any outliers or influential points, eliminate them and then repeat the variable selection procedures.

CHAPTER 9

Shrinkage Methods

9.1 Principal Components

Recall that if the X-matrix is orthogonal then testing and interpretation are greatly simplified. One motivation for principal components (PCs) is to rotate the X to orthogonality. For example, consider the case with two predictors depicted in Figure 9.1.

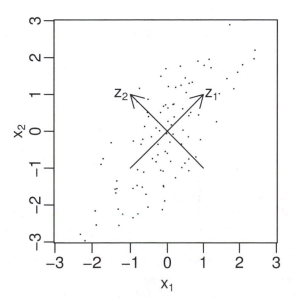

Figure 9.1 *Original predictors are x_1 and x_2; principal components are z_1 and z_2.*

The original predictors, x_1 and x_2, are clearly correlated and so the X-matrix will not be orthogonal. This will complicate the interpretation of the effects of x_1 and x_2 on the response. Suppose we rotate the coordinate axes so that in the new system, the predictors are orthogonal. Furthermore, suppose we make the rotation so that the first axis lies in the direction of the greatest variation in the data, the second in the second greatest direction of variation in those dimensions remaining and so on. These rotated directions, z_1 and z_2 in our two predictor example, are simply linear combinations of the original predictors. This is the geometrical description of PCs. We now indicate how these directions may be calculated.

We wish to find a rotation $p \times p$ matrix U such that $Z = XU$ and $Z^T Z = \text{diag}(\lambda_1, \cdots, \lambda_p)$ and $\lambda_1 \geq \lambda_2 \geq \cdots \geq \lambda_p \geq 0$. Zero eigenvalues indicate nonidentifiability. Since:

$$Z^T Z = U^T X^T X U$$

the eigenvalues of $X^T X$ are $\lambda_1, \ldots, \lambda_p$ and the eigenvectors of $X^T X$ are the columns of U. The columns of Z are called the PCs and these are orthogonal to each other. λ_i is the variance of Z_i.

Another way of looking at it is to try to find the linear combinations of X that have the maximum variation. We find the u_1 such that var $(u_1^T X)$ is maximized subject to $u_1^T u_1 = 1$. Now find u_2 such that var $(u_2^T X)$ is maximized subject to $u_1^T u_2 = 0$ and $u_2^T u_2 = 1$. We keep finding directions of greatest variation orthogonal to those directions we have already found. A simpler version of this is illustrated in Figure 9.1.

There are some variations on this theme:

1. We can use an X that includes the intercept as a column of ones or we can center each variable by its mean in which case we dispense with the intercept since the centered variables are orthogonal to it. The latter choice is more common since it means the PCs will just be linear combinations of predictors without a constant, which is easier to interpret.

2. As well as centering each variable, we could also divide by its standard deviation, thus standardizing each variable. This is equivalent to doing principal components analysis (PCA) on the correlation matrix of the predictors. This option makes the most sense when the predictors are measured on different scales, say millimeters and grams. If no scaling is done, the PCs will be dominated by the numerically largest predictors, but this may not make sense. When all the predictors are on the same scale, say distances between locations on the face, we might choose to center only.

PCA is a general technique for reducing the dimension of data. PCA is not specifically designed for regression and most applications lie in other fields. See Johnson and Wichern (2002) for an introduction. Typically, only a few eigenvalues will be large so that almost all the variation in X will be representable by the first few PCs. Various methods are used to determine how many components should be used. In application to regression, we replace the regression y ˜ X with y ˜ Z where we use only a few columns of Z. This is known as *principal components regression* or PCR. The technique is used in two distinct ways.

Explanation When the goal of the regression is to find simple, well-fitting and understandable models for the response, PCR *may* help. The PCs are linear combinations of the predictors. If we use these directly, then little is gained, but perhaps a close approximation may have a good interpretation. For example, suppose:

$$Z_1 = 0.71X_1 + 0.02X_2 - 0.06X_3 - 0.69X_4$$

Now Z_1 is approximately proportional to $X_1 - X_4$. So, suppose X_1 was a person's height and X_4 was the distance between his or her hands when outstretched to the side, this would represent a measure of the shape of the person. However, if X_4 was the person's weight, then $X_1 - X_4$ would be harder to interpret. If this interpretation process works well, a large number of X is reduced to a much smaller number of (approximate) Z that can be understood and used to model the response. However, for this to work, we typically need the predictors to measure quantities for which linear combinations are interpretable — usually the predictors would need the same units. Furthermore, we would need some luck to get interpretable PCs and we would need to make some creative approximations. These requirements severely restrict the utility of PCR for explanatory purposes.

Prediction It may be that we can make better predictions with a small number of Z than a much larger number of X. We shall see this in the example that follows. Success requires that we make a good choice of the number of components. PCA does not use y so it is possible (although less likely) that some lesser PC is actually very important in predicting the response.

A Tecator Infratec Food and Feed Analyzer working in the wavelength range of 850 to 1050 nm by the near infrared transmission (NIT) principle was used to collect data on samples of finely chopped pure meat. 215 samples were measured. For each sample, the fat content was measured along with a 100 channel spectrum of absorbances. Since determining the fat content via analytical chemistry is time consuming, we would like to build a model to predict the fat content of new samples using the 100 absorbances which can be measured more easily. See Thodberg (1993).

The true performance of any model is hard to determine based on just the fit to the available data. We need to see how well the model does on new data not used in the construction of the model. For this reason, we will partition the data into two parts — a *training sample* consisting of the first 172 observations that will be used to build the models and a *testing sample* of the remaining 43 observations.

Let's start with the least squares fit:

```
> data(meatspec)
> model1 <- lm(fat ~ ., meatspec[1:172,])
> summary(model1)$r.squared
[1] 0.99702
```

We see that the fit of this model is already very good in terms of R^2. How well does this model do in predicting the observations in the test sample? We need a measure of performance — we use root mean square error (RMSE):

$$\sqrt{\left(\sum_{i=1}^{n} (\hat{y}_i - y_i)^2 / n \right)}$$

where $n = 43$ in this instance. We find for the training sample that:

```
> rmse <- function(x,y) sqrt(mean((x-y)^2))
> rmse(model1$fit,meatspec$fat[1:172])
[1] 0.69032
```

while for the test sample:

```
> rmse(predict(model1,meatspec[173:215,]),meatspec$fat[173:215])
[1] 3.814
```

We see that the performance is much worse for the test sample. This is not unusual, as the fit to the data we have almost always gives an overoptimistic sense of how well the model will do with future data. In this case, the actual error is about five times greater than the model itself suggests.

Now, it is quite likely that not all 100 predictors are necessary to make a good prediction. In fact, some of them might just be adding noise to the prediction and we could improve matters by eliminating some of them. We use the default stepwise model selection:

```
> model2 <- step(model1)
> rmse(model2$fit,meatspec$fat[1:172])
[1] 0.7095
> rmse(predict(model2,meatspec[173:215,]),meatspec$fat[173:215])
[1] 3.5902
```

The model selection step removed 28 variables. Of course, the nominal fit got a little worse as it always will when predictors are removed, but the actual performance improved somewhat from 3.81 to 3.59.

Now let's compute the PCA on the training sample predictors:

```
> meatpca <- prcomp(meatspec[1:172,-101])
```

We can examine the square roots of the eigenvalues:

```
> round(meatpca$sdev,3)
 [1] 5.055 0.511 0.282 0.168 0.038 0.025 0.014 0.011 ...
[97] 0.000 0.000 0.000 0.000
```

It is better to examine the square roots because the eigenvalues themselves are the variances of the PCs, but standard deviations are a better scale for comparison. We see that the first PC accounts for about ten times more variation than the second. The contribution drops off sharply. This suggests that most of the variation in the predictors can be explained with just a few dimensions.

The eigenvectors can be found in the object meatpca$rotation. We plot these vectors against the predictor number (which represents a range of frequencies in this case) in Figure 9.2 using:

```
> matplot(1:100,meatpca$rot[,1:3],type="l",
  xlab="Frequency",ylab="")
```

These vectors are used for the linear combinations of the predictors that generate the PCs. We see that the first PC comes from an almost constant combination of the frequencies. It measures whether the predictors are generally large or small. The second PC represents a contrast between the higher and lower frequencies. The third is more difficult to interpret. It is sometimes possible, as in this example, to give some meaning to the PCs. This is typically a matter of intuitive interpretation. In some other cases, no interpretation can be found — this is almost always the case when the predictors measure variables on different scales (like a person's height and age).

We can get the PCs themselves from the columns of the object meatpca$x. Let's use the first four PCs to predict the response:

```
> model3 <- lm(fat ~ meatpca$x[,1:4]  , meatspec[1:172,])
> rmse(model3$fit,meatspec$fat[1:172])
[1] 4.0647
```

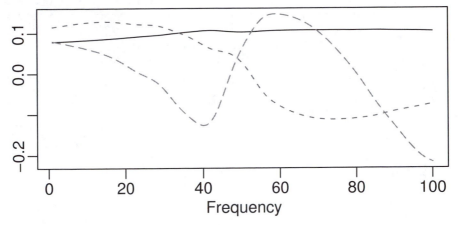

Figure 9.2 *Eigenvectors for the PCA of the meat spectrometer data. The solid line corresponds to the first PC, the dotted is for the second PC and the dashed is for the third PC.*

We do not expect as good a fit using only four variables instead of the 100. Even so, considering that, the fit is not much worse than the much bigger models.

PCR is an example of *shrinkage* estimation. Let's see where the name comes from. We plot the 100 slope coefficients for the full least squares fit:

```
> plot(model1$coef[-1],ylab="Coefficient")
```

which is shown in the left panel of Figure 9.3. We see that the coefficients range is in the thousands and that the adjacent coefficients can be very different. This is perhaps surprising because one might expect that adjacent frequencies might have a very similar effect on the response. Because the PCs represent a linear combination of the 100 predictors, we can compute the contribution of each in the PCR of just the four PCs and plot it. The PCR model is $y = Z\gamma + \varepsilon$ which is $y = XU\gamma + \varepsilon$. We compute $U\gamma$ and plot it in the panel on the right of Figure 9.3.

```
> svb <- meatpca$rot[,1:4] %*% model3$coef[-1]
> plot(svb,ylab="Coefficient")
```

Here we see that the coefficients are much smaller, ranging from -10 to 10 rather than in the thousands. Instead of wildly varying coefficients in the least squares case, we have a more stable result. This is why the effect is known as shrinkage. Furthermore, there is some smoothness between adjacent frequencies.

Why use four PCs here? The standard advice for choosing the number of PCs to represent the variation in X is to choose the number beyond which all the eigenvalues are relatively small. A good way to determine this number is to make a *scree plot*, which simply makes an index plot of the eigenvalues. We show here the square roots of the eigenvalues and only the first ten values to focus on the area of interest; see Figure 9.4:

```
> plot(meatpca$sdev[1:10],type="l",ylab="SD of PC",
  xlab="PC number")
```

We could make a case for using only the first PC, but there is another identifiable

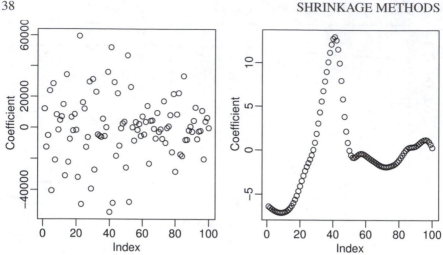

Figure 9.3 *Coefficients for the least squares fit on the left and for the PCR with four components on the right.*

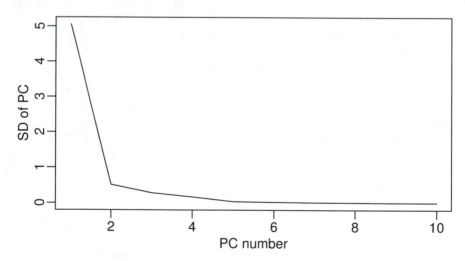

Figure 9.4 *Scree plot of the standard deviations of the first ten principal components.*

"elbow" at five indicating the choice of four PCs. Now let's see how well the test sample is predicted. The default version of PCs used here centers the predictors so we need to impose the same centering (using the means of the training sample) on the predictors:

```
> mm <- apply(meatspec[1:172,-101],2,mean)
> tx <- as.matrix(sweep(meatspec[173:215,-101],2,mm))
```

We now form the four linear combinations determined by the first four eigenvectors and compute the predicted values:

```
> nx <- tx %*% meatpca$rot[,1:4]
> pv <- cbind(1,nx) %*% model3$coef
```

We find the RMSE to be:

```
> rmse(pv,meatspec$fat[173:215])
[1] 4.534
```

which is not at all impressive. It turns out that we can do better by using more PCs — we figure out how many would give the best result on the test sample:

```
> rmsmeat <- numeric(50)
> for(i in 1:50){
+ nx <- tx %*% meatpca$rot[,1:i]
+ model3 <- lm(fat ~ meatpca$x[,1:i]  , meatspec[1:172,])
+ pv <- cbind(1,nx) %*% model3$coef
+ rmsmeat[i] <- rmse(pv,meatspec$fat[173:215])
+ }
> plot(rmsmeat,ylab="Test RMS",xlab="No. of Components")
> which.min(rmsmeat)
[1] 27
> min(rmsmeat)
[1] 1.8549
```

The plot of the RMSE is seen in Figure 9.5. The best result occurs for 27 PCs for which the RMSE is far better than anything achieved thus far. Of course, in practice we would not have access to the test sample in advance and so we would not know to use 27 components. We could, of course, reserve part of our original dataset for testing. This is sometimes called a *validation sample*. This is a reasonable strategy, but the downside is that we lose this sample from our estimation which degrades its quality. Furthermore, there is the question of which and how many observations should go into the validation sample. We can avoid this dilemma with the use of *crossvalidation* (CV). We divide the data into m parts, equal (or close to) in size. For each part, we use the rest of the data as the training set and that part as the test set. We evaluate the criterion of interest, RMSE in this case. We repeat for each part and average the result.

The `pls` package can compute this CV. By default, the data is divided into ten parts for the CV:

```
> library(pls)
> pcrmod <- pcr(fat ~ ., data=meatspec[1:172,], validation="CV",nc
> validationplot(pcrmod)
```

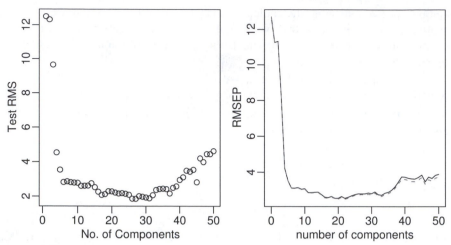

Figure 9.5 *RMS for the test sample on the left and RMS estimated using CV for varying numbers of PCs on the right.*

The crossvalidated estimates of the RMSE are shown in the right panel of Figure 9.5. The minimum occurs at 21 components. This gives an RMSE on the test sample, 2.21, that is close to the optimum.

9.2 Partial Least Squares

Partial least squares (PLS) is a method for relating a set of input variables $X_1, \ldots, X_m$ and outputs $Y_1, \ldots, Y_l$. PLS was developed by Herman Wold — see Wold, Ruhe, Wold, and Dunn (1984). PLS regression is comparable to PCR in that both predict the response using some number of linear combinations of the predictors. The difference is that while PCR ignores Y in determining the linear combinations, PLS regression explicitly chooses them to predict Y as well as possible.

We will consider only univariate PLS — that is to say $l = 1$ so that Y is scalar. We will attempt to find models of the form:

$$\hat{y} = \beta_1 T_1 + \cdots + \beta_p T_p$$

where T_k is a linear combination of the Xs. See Figure 9.6

Various algorithms have been presented for computing PLS. Most work by iteratively determining the T_is to predict y well, but at the same time maintaining orthogonality. One criticism of PLS is that it solves no well-defined modeling problem, which makes it difficult to distinguish between the competing algorithms on theoretical rather than empirical grounds. Garthwaite (1994) presents an intuitive algorithm, but de Jong (1993) describes the SIMPLS method, which is now the most well-known. Several other algorithms exist.

As with PCR, we must choose the number of components carefully. CV can be helpful in doing this. We apply PLS to the meat spectroscopy data using CV to select

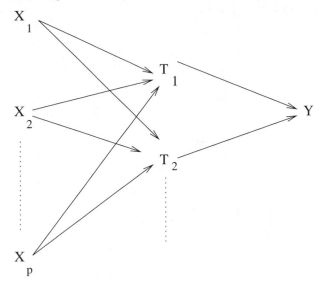

Figure 9.6 *Schematic representation of partial least squares.*

the number of components. We now compute the PLS on all models up to size 50. We plot the linear combination used for a four-component model in the left panel of Figure 9.7. The crossvalidated estimates of the RMSE are shown in the right panel.

```
> plsg <- plsr(fat ~ ., data=meatspec[1:172,], ncomp=50,
    validation="CV")
> coefplot(plsg,ncomp=4,xlab="Frequency")
> validationplot(plsg)
```

We see that the effective linear combination of the predictors in the four-component model is similar to the PCR indicating the shrinkage effect. As before, four components do not appear to be enough — we need around 14 components as suggested by the crossvalidated estimate of the RMSE. We need only half the number of components as PCR, which is expected since we are using information about the response. Note that the CV randomly divides the data into ten groups so if you repeat this calculation, you will not get exactly the same result.

Now we determine the performance on the training set for the 14-component model:

```
> ypred <- predict(plsg,ncomp=14)
> rmse(ypred,meatspec$fat[1:172])
[1] 1.9528
```

which is similar to PCR, but now see how we do on the test set. Because the response is centered by PLS we need to add the mean response for the training set back in. We also need to center the test set:

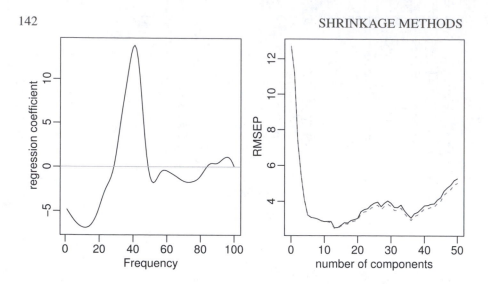

Figure 9.7 *Coefficients of X for a four-component PLS model are shown on the left. Estimated CV error is shown on the right.*

```
> ytpred <- predict(plsg,meatspec[173:215,],ncomp=14)
> rmse(ytpred,meatspec$fat[173:215])
[1] 2.0112
```

which is slightly better than the 2.21 achieved by PCR.

We have not checked any diagnostics in this analysis. PLS and PCR are just as sensitive to assumptions as OLS so these are still mandatory in any full analysis.

PCR and PLS compared

PCR and PLS have the biggest advantage over OLS when there are large numbers of variables relative to the number of cases. They do not even require that $n \geq p$. PCR attempts to find linear combinations of the predictors that explain most of the variation in these predictors using just a few components. The purpose is dimension reduction. Because the PCs can be linear combinations of all the predictors, the number of variables used is not always reduced. Because the PCs are selected using only the X-matrix and not the response, there is no definite guarantee that the PCR will predict the response particularly well, although it turns out that way. If it happens that we can interpret the PCs in a meaningful way, we may achieve a much simpler explanation of the response. Thus PCR is geared more towards explanation than prediction.

In contrast, PLS finds linear combinations of the predictors that best explain the response. It is most effective when there are large numbers of variables to be considered. If successful, the variability of prediction is substantially reduced. On the other hand, PLS is virtually useless for explanation purposes.

9.3 Ridge Regression

Ridge regression makes the assumption that the regression coefficients (after normalization) are not likely to be very large. The idea of shrinkage is therefore embedded in the method. It is appropriate for use when the design matrix is collinear and the usual least squares estimates of β appear to be unstable.

Suppose that the predictors have been centered by their means and scaled by their standard deviations and that the response has been centered. The ridge regression estimates of βs are then given by:

$$\hat{\beta} = (X^T X + \lambda I)^{-1} X^T y$$

The use of ridge regression can be motivated in two ways. Suppose we take a Bayesian point of view and put a prior (multivariate normal) distribution on β that expresses the belief that smaller values of β are more likely than larger ones. Large values of λ correspond to a belief that the β are really quite small whereas smaller values of λ correspond to a more relaxed belief about β. This is illustrated in Figure 9.8.

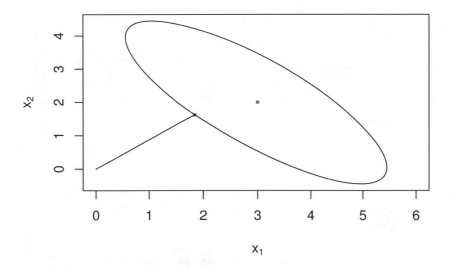

Figure 9.8 *Ridge regression illustrated. The least squares estimate is at the center of the ellipse while the ridge regression is the point on the ellipse closest to the origin. The ellipse is a contour of equal density of the posterior probability, which in this case will be comparable to a confidence ellipse. λ controls the size of the ellipse — the larger λ is, the larger the ellipse will be.*

Another way of looking at it is to suppose we place some upper bound on $\beta^T \beta$

and then compute the least squares estimate of β subject to this restriction. Use of Lagrange multipliers leads to ridge regression. The choice of λ corresponds to the choice of an upper bound in this formulation.

λ may be chosen by automatic methods, but it is also safer to plot the values of $\hat{\beta}$ as a function of λ. You should pick the smallest value of λ that produces stable estimates of β.

We demonstrate the method on the meat spectroscopy data; $\lambda = 0$ corresponds to least squares while we find that as $\lambda \to \infty$: $\hat{\beta} \to 0$.

```
> library(MASS)
> yc <- meatspec$fat[1:172]-mean(meatspec$fat[1:172])
> gridge <- lm.ridge(yc ~ trainx,lambda = seq(0,5e-8,1e-9))
> matplot(gridge$lambda,t(gridge$coef),type="l",lty=1,
   xlab=expression(lambda),ylab=expression(hat(beta)))
```

Some experimentation was necessary to determine the appropriate range of λ. The *ridge trace plot* is shown in Figure 9.9.

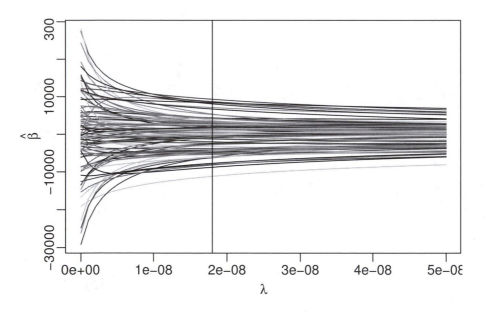

Figure 9.9 *Ridge trace plot for the meat spectroscopy data. The generalized crossvalidation choice of λ is shown as a vertical line.*

Various automatic selections for λ are available:

```
> select(gridge)
modified HKB estimator is 1.0583e-08
modified L-W estimator is 0.70969
smallest value of GCV  at 1.8e-08
> abline(v=1.8e-8)
```

We will use the generalized crossvalidation (GCV) estimate of $1.8e^{-8}$. First, we compute the training sample performance. This ridge regression both centers and scales the predictors, so we need to do the same in computing the fit. Furthermore, we need to add back in the mean of the response because of the centering:

```
> which.min(gridge$GCV)
1.8e-08
     19
> ypredg <- scale(trainx,center=FALSE,scale=gridge$scales)
    %*% gridge$coef[,19] + mean(meatspec$fat[1:172])
> rmse(ypredg,meatspec$fat[1:172])
[1] 0.80454
```

which is comparable to the above, but for the test sample we find:

```
> ytpredg <- scale(testx,center=FALSE,scale=gridge$scales)
    %*% gridge$coef[,19] + mean(meatspec$fat[1:172])
> rmse(ytpredg,meatspec$fat[173:215])
[1] 4.0966
```

which is dismayingly poor. However, a closer examination of the predictions reveals that just one of the ridge predictions is bad:

```
> c(ytpredg[13],ytpred[13],meatspec$fat[172+13])
     185      185
11.188 35.690 34.800
```

The PLS prediction (second) is close to the truth (third), but the ridge prediction is bad. If we remove this case:

```
> rmse(ytpredg[-13],meatspec$fat[173:215][-13])
[1] 1.9765
```

we get a good result.

Ridge regression estimates of coefficients are biased. Bias is undesirable, but it is not the only consideration. The mean-squared error (MSE) can be decomposed in the following way:

$$E(\hat{\beta} - \beta)^2 = (E(\hat{\beta} - \beta))^2 + E(\hat{\beta} - E\hat{\beta})^2$$

Thus the MSE of an estimate can be represented as the square of the bias plus the variance. Sometimes a large reduction in the variance may be obtained at the price of an increase in the bias. If the MSE is reduced as a consequence, then we may be willing to accept some bias. This is the trade-off that ridge regression makes — a reduction in variance at the price of an increase in bias. This is a common dilemma.

Frank and Friedman (1993) compared PCR, PLS and ridge regression and found the best results for ridge regression. Of course, for any given dataset any of the methods may prove to be the best, so picking a winner is difficult.

Exercises

1. Using the `seatpos` data, perform a PCR analysis with `hipcenter` as the response and `HtShoes`, `Ht`, `Seated`, `Arm`, `Thigh` and `Leg` as predictors. Select an appropriate number of components and give an interpretation to

those you choose. Add `Age` and `Weight` as predictors and repeat the analysis. Use both models to predict the response for predictors taking these values:

Age	Weight	HtShoes	Ht	Seated
64.800	263.700	181.080	178.560	91.440
Arm	Thigh	Leg		
35.640	40.950	38.790		

2. Fit a **PLS** model to the `seatpos` data with `hipcenter` as the response and all other variables as predictors. Take care to select an appropriate number of components. Use the model to predict the response at the values of the predictors specified in the first question.

3. Fit a ridge regression model to the `seatpos` data with `hipcenter` as the response and all other variables as predictors. Take care to select an appropriate amount of shrinkage. Use the model to predict the response at the values of the predictors specified in the first question.

4. Take the `fat` data, and use the percentage of body fat as the response and the other variables as potential predictors. Remove every tenth observation from the data for use as a test sample. Use the remaining data as a training sample building the following models:

   (a) Linear regression with all predictors
   (b) Linear regression with variables selected using AIC
   (c) Principal component regression
   (d) Partial least squares
   (e) Ridge regression

   Use the models you find to predict the response in the test sample. Make a report on the performance of the models.

CHAPTER 10

Statistical Strategy and Model Uncertainty

10.1 Strategy

Thus far we have learned various tactics:

1. *Diagnostics:* Checking of assumptions — constant variance, linearity, normality, outliers, influential points, serial correlation and collinearity

2. *Transformation:* Transforming the response — Box–Cox, transforming the predictors — splines and polynomial regression

3. *Variable selection:* Testing- and criterion-based methods

In what order should these be done? Should procedures be repeated at later stages? When should we stop? There are no definite answers to these questions. If the reader insists, I would recommend *Diagnostics → Transformation → Variable Selection → Diagnostics* as a rudimentary strategy. However, regression analysis is a search for structure in data and there are no hard-and-fast rules about how it should be done. Regression analysis requires some skill. You must be alert to unexpected structure in the data. Thus far, no one has implemented a computer program for conducting a complete analysis. Because of the difficulties in automating the assessment of regression graphics in an intelligent manner, I do not expect that this will be accomplished soon. The human analyst has the ability to assess plots in light of contextual information about the data.

There is a danger of doing too much analysis. The more transformations and permutations of leaving out influential points you do, the better fitting model you will find. Torture the data long enough, and sooner or later it will confess. Remember that fitting the data well is no guarantee of good predictive performance or that the model is a good representation of the underlying population. So:

1. Avoid complex models for small datasets.

2. Try to obtain new data to validate your proposed model. Some people set aside some of their existing data for this purpose.

3. Use past experience with similar data to guide the choice of model.

Data analysis is not an automatic process. Analysts have personal preferences in their choices of methodology, use software with varying capabilities and will interpret the same graphical display differently. In comparing the competing analyses of two statisticians, it may sometimes be possible to determine that one analysis is clearly superior. However, in most cases, particularly when the analysts are experienced, a universally acceptable judgment of superiority will not be possible.

The same data may support different models. Conclusions drawn from the models may differ quantitatively and qualitatively. However, except for those well-known datasets that circulate endlessly through textbooks and research articles, most data are only analyzed once. The analyst may be unaware that a second independent look at the data may result in quite different conclusions. We call this problem *model multiplicity*. In the next section, we describe an experiment illustrating the depth of this problem.

10.2 An Experiment in Model Building

In 1996, I taught a semester length masters level course in applied regression analysis to 28 students. Towards the end of the semester, I decided to set an assignment to test the students' ability in building a regression model for the purposes of prediction. I generated regression data with a response y and five uncorrelated predictors and $n = 50$ from a model known only to me, which was:

$$1/y = x_1 + 0.57x_1^2 + 4x_1x_2 + 2.1\exp(x_4) + \varepsilon$$

where $x_1 \sim U(0,1), x_2 \sim N(0,1), 1/x_3 \sim U(0,1), x_4 \sim N(1,1), x_5 \sim U(1,3)$ and $\varepsilon \sim N(0,1)$.

I asked students to predict the mean response at ten values of the predictors that I specified. I also asked them to provide a standard error for each of their predictions. The students understood and were reminded of the distinction between the standard error for the mean response and for a future observed value. The students were told that their score on the assignment would depend only on the closeness of their predicted values and the true values and on how closely their standard errors reflected the difference between these two quantities. Students were told to work independently.

For a given student's input, let p_i be his or her prediction, t_i be the true value and s_i be the standard error where $i = 1, \ldots, 10$. To assess his or her prediction accuracy, I used:

$$\sum_{i=1}^{10} \left(\frac{p_i - t_i}{t_i} \right)^2$$

whereas to measure the "honesty" of their standard errors, I used:

$$\frac{1}{10} \sum_{i=1}^{10} \left| \frac{p_i - t_i}{s_i} \right|$$

We would expect the predicted value to differ from the true value by typically about one standard error if the latter has been correctly estimated. Therefore, the measure of standard error honesty should be around one:

1.12	1.20	1.46	1.46	1.54	1.62	1.69
1.69	1.79	3.14	4.03	4.61	5.04	5.06
5.13	5.60	5.76	5.76	5.94	6.25	6.53
6.53	6.69	10.20	34.45	65.53	674.98	37285.95

Table 10.1 *Measures of prediction accuracy for 28 students in ascending order of magnitude.*

The prediction accuracy scores for the 28 students are shown in Table 10.1. We see that one student did very poorly. An examination of his or her model and some conversation revealed that this student neglected to back transform predictions to the original scale when using a model with a transform on the response. Three pairs of scores are identical in the table, but an examination of the models used and more significant digits revealed that only one pair was due to the students using the same model. This pair of students were known associates. Thus 27 different models were found by 28 students.

The scores for honesty of standard errors are shown in Table 10.2. The order in which scores are shown correspond to that given in Table 10.1:

0.75	7.87	6.71	0.59	4.77	8.20	11.74
10.70	1.04	17.10	3.23	14.10	84.86	15.52
80.63	17.61	14.02	14.02	13.35	16.77	12.15
12.15	12.03	68.89	101.36	18.12	2.24	40.08

Table 10.2 *Honesty of standard errors — order of scores corresponds to that in Table 10.1.*

We see that the students' standard errors were typically around an order of magnitude smaller than they should have been.

10.3 Discussion

Why was there so much model multiplicity? The students were all in the same class and used the same software, but almost everyone chose a different model. The course covered many of the commonly understood techniques for variable selection, transformation and diagnostics including outlier and influential point detection. The students were confronted with the problem of selecting the order in which to apply these methods and choosing from several competing methods for a given purpose.

The reason the models were so different was that students applied the various methods in different orders. Some did variable selection before transformation and others, the reverse. Some repeated a method after the model was changed and others did not. I went over the strategies that several of the students used and could not find anything clearly wrong with what they had done. One student made a mistake in computing his or her predicted values, but there was nothing obviously wrong in the remainder. The performance on this assignment did not show any relationship with that in the exams.

The implications for statistical practice are serious. Often a dataset is analyzed by a single analyst who comes up with a single model. Predictions and inferences are based on this single model. The analyst may be unaware that the data support quite different models which may lead to very different conclusions. Clearly one will not always have a stable of 28 independent analysts to search for alternatives, but it does point to the value of a second or third independent analysis. It may also be possible to automate the components of the analysis to some extent as in Faraway (1994) to see whether changes in the order of analysis might result in a different model.

Another issue is raised by the standard error results. Often we use the data to help determine the model. Once a model is built or selected, inferences and predictions may be made. Usually inferences are based on the assumption that the selected model was fixed in advance and so only reflect uncertainty concerning the parameters of that model. Students took that approach here. Because the uncertainty concerning the model itself is not allowed for, these inferences tend to be overly optimistic leading to unrealistically small standard errors. Methods for realistic inference when the data are used to select the model have come under the heading of *Model Uncertainty* — see Chatfield (1995) for a review. The effects of model uncertainty often overshadow the parametric uncertainty and the standard errors need to be inflated to reflect this. Faraway (1992) developed a bootstrap approach to compute these standard errors while Draper (1995) is an example of a Bayesian approach. These methods are a step in the right direction in that they reflect the uncertainty in model selection. Nevertheless, they do not address the problem of model multiplicity since they proscribe a particular method of analysis that does not allow for differences between human analysts.

Sometimes the data speak with a clear and unanimous voice — the conclusions are incontestable. Other times, differing conclusions may be drawn depending on the model chosen. We should acknowledge the possibility of alternative conflicting models and seek them.

CHAPTER 11

Insurance Redlining — A Complete Example

In this chapter, we present a relatively complete data analysis. The example is interesting because it illustrates several of the ambiguities and difficulties encountered in statistical practice.

Insurance redlining refers to the practice of refusing to issue insurance to certain types of people or within some geographic area. The name comes from the act of drawing a red line around an area on a map. Now few would quibble with an insurance company refusing to sell auto insurance to a frequent drunk driver, but other forms of discrimination would be unacceptable.

In the late 1970s, the U.S. Commission on Civil Rights examined charges by several Chicago community organizations that insurance companies were redlining their neighborhoods. Because comprehensive information about individuals being refused homeowners insurance was not available, the number of FAIR plan policies written and renewed in Chicago by zip code for the months of December 1977 through May 1978 was recorded. The FAIR plan was offered by the city of Chicago as a default policy to homeowners who had been rejected by the voluntary market. Information on other variables that might affect insurance writing such as fire and theft rates were also collected at the zip code level. The variables are:

race racial composition in percentage of minority

fire fires per 100 housing units

theft theft per 1000 population

age percentage of housing units built before 1939

involact new FAIR plan policies and renewals per 100 housing units

income median family income in thousands of dollars

side North or South Side of Chicago

The data come from Andrews and Herzberg (1985) where more details of the variables and the background are provided.

11.1 Ecological Correlation

Notice that we do not know the race of those denied insurance. We only know the racial composition in the corresponding zip code. This is an important difficulty that needs to be considered before starting the analysis.

When data are collected at the group level, we may observe a correlation between two variables. The ecological fallacy is concluding that the same correlation holds at the individual level. For example, in countries with higher fat intakes in the diet, higher rates of breast cancer have been observed. Does this imply that individuals with high fat intakes are at a higher risk of breast cancer? Not necessarily. Relationships seen in observational data are subject to confounding, but even if this is allowed for, bias is caused by aggregating data. We consider an example taken from U.S. demographic data:

```
> data(eco)
> plot(income ~ usborn, data=eco, xlab="Proportion US born",
  ylab="Mean Annual Income")
```

In the first panel of Figure 11.1, we see the relationship between 1998 per capita income dollars from all sources and the proportion of legal state residents born in the United States in 1990 for each of the 50 states plus the District of Columbia (D.C.). We can see a clear negative correlation.

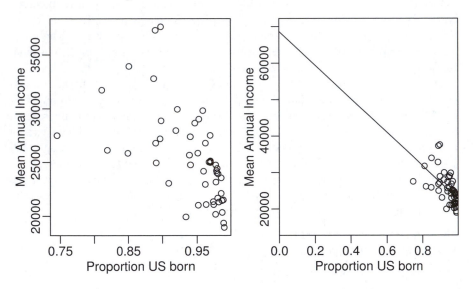

Figure 11.1 *1998 annual per capita income and proportion U.S. born for 50 states plus D.C. The plot on the right shows the same data as on the left, but with an extended scale and the least squares fit shown.*

We can fit a regression line and show the fitted line on an extended range:

```
> g <- lm(income ~ usborn, eco)
> summary(g)
Coefficients:
            Estimate Std. Error t value Pr(>|t|)
(Intercept)    68642       8739    7.85  3.2e-10
usborn        -46019       9279   -4.96  8.9e-06

Residual standard error: 3490 on 49 degrees of freedom
Multiple R-Squared: 0.334,      Adjusted R-squared: 0.321
F-statistic:  24.6 on 1 and 49 DF,  p-value: 8.891e-06
```

```
> plot(income ~ usborn, data=eco, xlab="Proportion US born",
  ylab="Mean Annual Income",xlim=c(0,1),ylim=c(15000,70000),
  xaxs="i")
> abline(coef(g))
```

We see that there is a clear statistically significant relationship between the per capita annual income and the proportion who are U.S. born. What does this say about the average annual income of people who are U.S. born and those who are naturalized citizens? If we substitute `usborn=1` into the regression equation, we get 68642 − 46019 = $22,623, while if we put `usborn=0`, we get $68,642. This suggests that on average, naturalized citizens earn three times more than U.S. born citizens. In truth, information from the U.S. Bureau of the Census indicates that U.S. born citizens have an average income just slightly larger than naturalized citizens. What went wrong with our analysis?

The ecological inference from the aggregate data to the individuals requires an assumption of constancy. Explicitly, the assumption would be that the incomes of the native born do not depend on the proportion of native born within the state (and similarly for naturalized citizens). This assumption is unreasonable for these data because immigrants are naturally attracted to wealthier states.

This assumption is also relevant to the analysis of the Chicago Insurance data since we have only aggregate data. We must keep in mind that the results for the aggregated data may not hold true at the individual level.

11.2 Initial Data Analysis

Start by reading the data in and examining it:

```
> data(chredlin)
> chredlin
      race fire theft  age involact income side
60626 10.0  6.2    29 60.4      0.0 11.744    n
60640 22.2  9.5    44 76.5      0.1  9.323    n
...etc...
60645  3.1  4.9    27 46.6      0.0 13.731    n
```

Summarize:

```
> summary(chredlin)
      race              fire              theft             age
 Min.   : 1.00    Min.   : 2.00    Min.   :   3.0    Min.   : 2.0
 1st Qu.: 3.75    1st Qu.: 5.65    1st Qu.:  22.0    1st Qu.:48.6
 Median :24.50    Median :10.40    Median :  29.0    Median :65.0
 Mean   :34.99    Mean   :12.28    Mean   :  32.4    Mean   :60.3
 3rd Qu.:57.65    3rd Qu.:16.05    3rd Qu.:  38.0    3rd Qu.:77.3
 Max.   :99.70    Max.   :39.70    Max.   : 147.0    Max.   :90.1
    involact            income           side
 Min.   :0.000    Min.   : 5.58    n:25
 1st Qu.:0.000    1st Qu.: 8.45    s:22
 Median :0.400    Median :10.69
```

```
Mean    :0.615    Mean    :10.70
3rd Qu.:0.900    3rd Qu.:11.99
Max.    :2.200    Max.    :21.48
```

We see that there is a wide range in the `race` variable, with some zip codes almost entirely minority or nonminority. This is good for our analysis since it will reduce the variation in the regression coefficient for race, allowing us to assess this effect more accurately. If all the zip codes were homogeneous, we would never be able to discover an effect from these aggregated data. We also note some skewness in the `theft` and `income` variables. The response `involact` has a large number of zeros. This is not good for the assumptions of the linear model but we have little choice but to proceed. We will not use the information about North vs. South Side until later. Now make some graphical summaries:

```
> par(mfrow=c(2,3))
> for(i in 1:6) stripchart(chredlin[,i],main=names(chredlin)[i],
  vertical=TRUE,method="jitter")
> par(mfrow=c(1,1))
> pairs(chredlin)
```

The strip plots are seen in Figure 11.2. Jittering has been added to avoid overplotting of symbols. Now look at the relationship between `involact` and `race`:

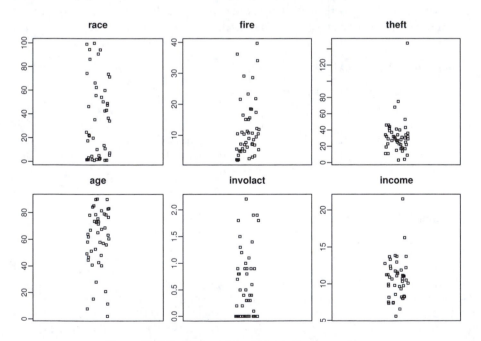

Figure 11.2 *Strip plots of the Chicago Insurance data.*

```
> summary(lm(involact ~ race,chredlin))
Coefficients:
```

```
             Estimate Std. Error t value Pr(>|t|)
(Intercept)   0.12922    0.09661    1.34     0.19
race          0.01388    0.00203    6.84  1.8e-08
```

```
Residual standard error: 0.449 on 45 degrees of freedom
Multiple R-Squared: 0.509,       Adjusted R-squared: 0.499
F-statistic: 46.7 on 1 and 45 DF,  p-value: 1.78e-08
```

We can clearly see that homeowners in zip codes with a high percentage of minorities are taking the default FAIR plan insurance at a higher rate than other zip codes. That is not in doubt. However, can the insurance companies claim that the discrepancy is due to greater risks in some zip codes? The insurance companies could claim that they were denying insurance in neighborhoods where they had sustained large fire-related losses and any discriminatory effect was a by-product of legitimate business practice. We plot some of the variables involved by this question in Figure 11.3:

```
> plot(involact ~ race, chredlin)
> abline(lm(involact ~ race, chredlin))
> plot(fire ~ race, chredlin)
> abline(lm(fire ~ race, chredlin))
```

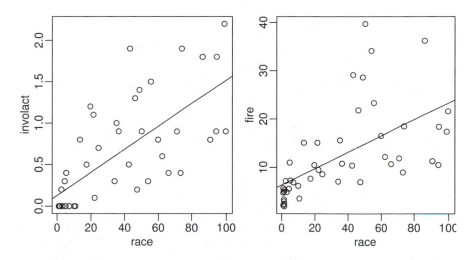

Figure 11.3 *Relationship between* fire, race *and* involact *in the Chicago data.*

The question of which variables should also be included in the regression so that their effect may be adjusted for is difficult. Statistically, we can do it, but the important question is whether it should be done at all. For example, it is known that the incomes of women in the United States and other countries are generally lower than those of men. However, if one adjusts for various factors such as type of job and length of service, this gender difference is reduced or can even disappear. The controversy is not statistical but political — should these factors be used to make the adjustment?

For the present data, suppose that the effect of adjusting for income differences was to remove the race effect. This would pose an interesting, but nonstatistical question. I have chosen to include the `income` variable in the analysis just to see what happens.

I have decided to use `log(income)` partly because of skewness in this variable, but also because income is better considered on a multiplicative rather than additive scale. In other words, $1,000 is worth a lot more to a poor person than a millionaire because $1,000 is a much greater fraction of the poor person's wealth.

11.3 Initial Model and Diagnostics

We start with the full model:

```
> g <- lm(involact ~ race + fire + theft + age + log(income),
  chredlin)
> summary(g)
Coefficients:
              Estimate Std. Error t value Pr(>|t|)
(Intercept) -1.18554    1.10025   -1.08   0.28755
race         0.00950    0.00249    3.82   0.00045
fire         0.03986    0.00877    4.55   4.8e-05
theft       -0.01029    0.00282   -3.65   0.00073
age          0.00834    0.00274    3.04   0.00413
log(income)  0.34576    0.40012    0.86   0.39254

Residual standard error: 0.335 on 41 degrees of freedom
Multiple R-Squared: 0.752,       Adjusted R-squared: 0.721
F-statistic: 24.8 on 5 and 41 DF,  p-value: 2.01e-11
```

Before leaping to any conclusions, we should check the model assumptions. These two diagnostic plots are seen in Figure 11.4:

```
> plot(fitted(g),residuals(g),xlab="Fitted",ylab="Residuals")
> abline(h=0)
> qqnorm(residuals(g))
> qqline(residuals(g))
```

The diagonal streak in the residual-fitted plot is caused by the large number of zero response values in the data. When $y = 0$, the residual $\hat{\varepsilon} = -\hat{y} = -x^T \hat{\beta}$, hence the line. Turning a blind eye to this feature, we see no particular problem. The Q–Q plot looks fine too.

Now let's look at influence — what happens if points are excluded? We plot the leave-out-one differences in $\hat{\beta}$ for `theft` and the Cook distances:

```
> gi <- influence(g)
> qqnorml(gi$coef[,4])
> halfnorm(cooks.distance(g))
```

See Figure 11.5 where cases 6 and 24 stick out. It is worth looking at other leave-out-one coefficient plots also. We check the jackknife residuals for outliers:

```
> range(rstudent(g))
[1] -3.1850  2.7929
```

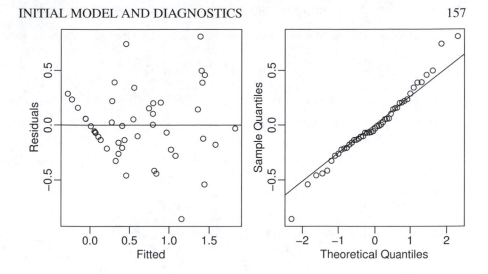

Figure 11.4 *Diagnostic plots of the initial model for the Chicago Insurance data.*

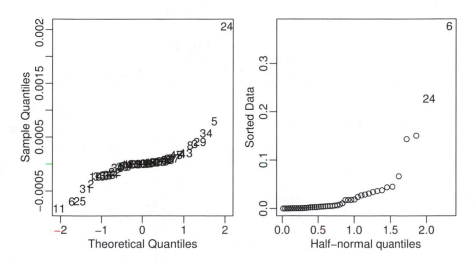

Figure 11.5 *A Q–Q plot of the leave-out-one coefficient differences for the* theft *variable is shown on the left. A half-normal plot of the Cook distances is shown on the right.*

There is nothing extreme enough to call an outlier. Let's take a look at the two cases:

```
> chredlin[c(6,24),]
      race fire theft  age involact income side
60610 54.0 34.1    68 52.6      0.3  8.231    n
60607 50.2 39.7   147 83.0      0.9  7.459    n
```

These are high theft and fire zip codes. See what happens when we exclude these points:

```
> g <- lm(involact ~ race + fire + theft + age + log(income),
  chredlin,subset=-c(6,24))
> summary(g)
Coefficients:
            Estimate Std. Error t value Pr(>|t|)
(Intercept) -0.57674    1.08005   -0.53    0.596
race         0.00705    0.00270    2.62    0.013
fire         0.04965    0.00857    5.79   1e-06
theft       -0.00643    0.00435   -1.48    0.147
age          0.00517    0.00289    1.79    0.082
log(income)  0.11570    0.40111    0.29    0.775

Residual standard error: 0.303 on 39 degrees of freedom
Multiple R-Squared: 0.804,       Adjusted R-squared: 0.779
F-statistic:   32 on 5 and 39 DF,  p-value: 8.2e-13
```

theft and age are no longer significant at the 5% level.

11.4 Transformation and Variable Selection

We now look for transformations. We try some partial residual plots as seen in Figure 11.6:

```
> prplot(g,1)
> prplot(g,2)
```

These plots indicate no need to transform. It would have been inconvenient to transform the race variable since that would have made interpretation more difficult. Fortunately, we do not need to worry about this. We examined the other partial residual plots and experimented with polynomials for the predictors. No transformation of the predictors appears to be worthwhile.

We choose to avoid even considering a transformation of the response. The zeros in the response would have restricted the possibilities and furthermore would have made interpretation more difficult.

We now move on to variable selection. We are not so much interested in picking one model here because we are mostly interested in the dependency of involact on the race variable. So $\hat{\beta}_1$ is the estimate we want to focus on. The problem is that collinearity with the other variables may cause $\hat{\beta}_1$ to vary substantially depending on what other variables are in the model. We address this question here. We leave out the two influential points and force race to be included in every model. We do this because race is the primary predictor of interest in this model and we want to

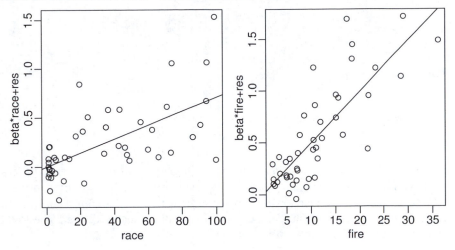

Figure 11.6 *Partial residual plots for* race *and* fire.

measure its effect. We are not prejudging whether it is significant or not — we will check that later:

```
> chreduc <- chredlin[-c(6,24),]
> library(leaps)
> b<-regsubsets(involact~race + fire + theft + age + log(income),
  force.in=1,data=chreduc)
> (rs <- summary(b))
Subset selection object
              Forced in Forced out
race                TRUE      FALSE
fire               FALSE      FALSE
theft              FALSE      FALSE
age                FALSE      FALSE
log(income)        FALSE      FALSE
1 subsets of each size up to 5
Selection Algorithm: exhaustive
          race fire theft age log(income)
2  ( 1 ) "*"  "*"  " "   " " " "
3  ( 1 ) "*"  "*"  " "   "*" " "
4  ( 1 ) "*"  "*"  "*"   "*" " "
5  ( 1 ) "*"  "*"  "*"   "*" "*"
> rs$adj
[1] 0.76855 0.77650 0.78402 0.77895
```

The best model seems to be this one:

```
> g <- lm(involact ~ race + fire + theft + age, chredlin,
  subset=-c(6,24))
> summary(g)
Coefficients:
```

```
             Estimate Std. Error t value Pr(>|t|)
(Intercept) -0.26787   0.13967   -1.92    0.0623
race         0.00649   0.00184    3.53    0.0011
fire         0.04906   0.00823    5.96   5.3e-07
theft       -0.00581   0.00373   -1.56    0.1271
age          0.00469   0.00233    2.01    0.0514
```

```
Residual standard error: 0.3 on 40 degrees of freedom
Multiple R-Squared: 0.804,       Adjusted R-squared: 0.784
F-statistic: 40.9 on 4 and 40 DF,  p-value: 1.24e-13
```

The fire rate is significant and actually has higher t-statistics; but nevertheless, we have verified that there is a positive relationship between involact and race while controlling for a selection of the other variables. Even so we must consider the reliability of this conclusion. For example, would other analysts have come to the same conclusion? One alternative model is:

```
> galt <- lm(involact ~ race+fire+log(income),chredlin,
  subset=-c(6,24))
> summary(galt)
Coefficients:
             Estimate Std. Error t value Pr(>|t|)
(Intercept)  0.75326   0.83588    0.90     0.373
race         0.00421   0.00228    1.85     0.072
fire         0.05102   0.00845    6.04   3.8e-07
log(income) -0.36238   0.31916   -1.14     0.263
```

```
Residual standard error: 0.309 on 41 degrees of freedom
Multiple R-Squared: 0.786,       Adjusted R-squared: 0.77
F-statistic: 50.1 on 3 and 41 DF,  p-value: 8.87e-14
```

In this model, we see that race is not statistically significant. The previous model did fit slightly better, but it is important that there exists a reasonable model in which race is not significant since, although the evidence seems fairly strong in favor of a race effect, it is not entirely conclusive. Interestingly enough, if log(income) is now dropped:

```
> galt <- lm(involact ~ race+fire,chredlin,subset=-c(6,24))
> summary(galt)
Coefficients:
             Estimate Std. Error t value Pr(>|t|)
(Intercept) -0.19132   0.08152   -2.35    0.0237
race         0.00571   0.00186    3.08    0.0037
fire         0.05466   0.00784    6.97   1.6e-08
```

```
Residual standard error: 0.31 on 42 degrees of freedom
Multiple R-Squared: 0.779,       Adjusted R-squared: 0.769
F-statistic: 74.1 on 2 and 42 DF,  p-value: 1.70e-14
```

we find race again becomes significant, which raises again the question of whether income should be adjusted for since it makes all the difference here.

We now return to the two left-out cases. Observe the difference in the fit when the two are reincluded on the best model. The quantities may change but the qualitative message is the same. It is better to include all points if possible, especially in a legal case like this, where excluding points might lead to criticism and suspicion of the results:

```
> g <- lm(involact ~ race + fire + theft + age, chredlin)
> summary(g)
Coefficients:
             Estimate Std. Error t value Pr(>|t|)
(Intercept) -0.24312    0.14505   -1.68   0.10116
race         0.00810    0.00189    4.30   0.00010
fire         0.03665    0.00792    4.63   3.5e-05
theft       -0.00959    0.00269   -3.57   0.00092
age          0.00721    0.00241    2.99   0.00460

Residual standard error: 0.334 on 42 degrees of freedom
Multiple R-Squared: 0.747,      Adjusted R-squared: 0.723
F-statistic:   31 on 4 and 42 DF,  p-value: 4.8e-12
```

The main message of the data is not changed. On checking the diagnostics, I found no trouble. So it looks like there is moderately good evidence that zip codes with high minority populations are being "redlined." While there is evidence that some of the relationship between `race` and `involact` can be explained by the fire rate, there is still a component that cannot be attributed to the other variables.

11.5 Discussion

There is some ambiguity in the conclusion here. These reservations have several sources.

There is some doubt because the response is not a perfect measure of people being denied insurance. It is an aggregate measure that raises the problem of ecological correlations. We have implicitly assumed that the probability a minority homeowner would obtain a FAIR plan after adjusting for the effect of the other covariates is constant across zip codes. This is unlikely to be true. If the truth is simply a variation about some constant, then our conclusions will still be reasonable, but if this probability varies in a systematic way, then our conclusions may be off the mark. It would be a very good idea to obtain some individual level data.

Another point to be considered is the size of the effect. The largest value of the response is only 2.2% and most other values are much smaller. Even assuming the worst, the number of people affected is small.

There is also the problem of a potential latent variable that might be the true cause of the observed relationship. Someone with firsthand knowledge of the insurance business might propose one. This possibility always casts a shadow of doubt on our conclusions.

Another issue that arises in cases of this nature is how much the data should be aggregated. For example, suppose we fit separate models to the two halves of the city. Fit the model to the south of Chicago:

```
> g <- lm(involact ~ race+fire+theft+age, subset=(side == "s"),
  chredlin)
> summary(g)
Coefficients:
              Estimate Std. Error t value Pr(>|t|)
(Intercept)  -0.23441    0.23774   -0.99    0.338
race          0.00595    0.00328    1.81    0.087
fire          0.04839    0.01689    2.87    0.011
theft        -0.00664    0.00844   -0.79    0.442
age           0.00501    0.00505    0.99    0.335

Residual standard error: 0.351 on 17 degrees of freedom
Multiple R-Squared: 0.743,        Adjusted R-squared: 0.683
F-statistic: 12.3 on 4 and 17 DF,  p-value: 6.97e-05
```

and now to the north:

```
> g <- lm(involact ~ race+fire+theft+age, subset=(side == "n"),
  chredlin)
> summary(g)
Coefficients:
              Estimate Std. Error t value Pr(>|t|)
(Intercept)  -0.31857    0.22702   -1.40    0.176
race          0.01256    0.00448    2.81    0.011
fire          0.02313    0.01398    1.65    0.114
theft        -0.00758    0.00366   -2.07    0.052
age           0.00820    0.00346    2.37    0.028

Residual standard error: 0.343 on 20 degrees of freedom
Multiple R-Squared: 0.756,        Adjusted R-squared: 0.707
F-statistic: 15.5 on 4 and 20 DF,  p-value: 6.52e-06
```

We see that race is significant in the north, but not in the south. By dividing the data into smaller and smaller subsets it is possible to dilute the significance of any predictor. On the other hand, it is important not to aggregate all data without regard to whether it is reasonable. Clearly a judgment has to be made and this can be a point of contention in legal cases.

There are some special difficulties in presenting this during a court case. With scientific inquiries, there is always room for uncertainty and subtlety in presenting the results, particularly if the subject matter is not contentious. In an adversarial proceeding, it is difficult to present statistical evidence when the outcome is not clear-cut, as in this example. There are particular difficulties in explaining such evidence to nonmathematically trained people.

After all this analysis, the reader may be feeling somewhat dissatisfied. It seems we are unable to come to any truly definite conclusions and everything we say has been hedged with "ifs" and "buts." Winston Churchill once said:

> Indeed, it has been said that democracy is the worst form of Government except all those other forms that have been tried from time to time.

We might say the same about statistics with respect to how it helps us reason in the face of uncertainty. It is not entirely satisfying but the alternatives are worse.

CHAPTER 12

Missing Data

Missing data occur when some values of some cases are missing. This is not uncommon. Dealing with missing data is time consuming. Fixing up problems caused by missing data sometimes takes longer than the analysis.

What can be done? Obviously, finding the missing values is the best option, but this is not always possible. Next, ask why the data are missing. If the reason an observation is missing is noninformative, then a fix is easier. For example, if a data point is missed because it was large in value, then this could cause some bias and a simple fix is not possible. Patients may drop out of a drug study, because they believe their treatment is not working — this would cause bias.

Here are several fix-up methods to use when data are missing for noninformative reasons:

1. Delete the case with missing observations. This is OK if this only causes the loss of a relatively small number of cases. This is the simplest solution.

2. Fill in or *impute* the missing values. Use the rest of the data to predict the missing values. Simply replacing the missing value of a predictor with the average value of that predictor is one easy method. Using regression on the other predictors is another possibility. It is not clear how much the diagnostics and inference on the filled-in dataset are affected. Some additional uncertainty is caused by the imputation, which needs to be taken into account. Multiple imputation can capture some of this uncertainty.

3. Consider just (x_i, y_i) pairs with some observations missing. The means and SDs of x and y can be used in the estimate even when a member of a pair is missing. An analogous method is available for regression problems. This is called the *missing value correlation* method.

4. Maximum likelihood methods can be used assuming the multivariate normality of the data. The EM algorithm is often used here. We will not explain the details, but the idea is essentially to treat missing values as nuisance parameters.

Suppose some of the values in the Chicago Insurance dataset were missing. I randomly declared some of the observations missing in this modified dataset. Read it in and take a look:

```
> data(chmiss)
> chmiss
      race fire theft  age involact income
60626 10.0  6.2    29 60.4       NA 11.744
60640 22.2  9.5    44 76.5      0.1  9.323
60613 19.6 10.5    36   NA      1.2  9.948
```

```
60657 17.3  7.7     37    NA        0.5 10.656
--- etc ---
60645  3.1  4.9     27    NA        0.0 13.731
```

There are 20 missing observations denoted by NA here. It is important to know what the missing value code is for the data and/or software you are using. See what happens if we try to fit the model:

```
> g <- lm(involact ~ .,chmiss)
> summary(g)
Coefficients:
             Estimate Std. Error t value Pr(>|t|)
(Intercept) -1.11648    0.60576   -1.84  0.07947
race         0.01049    0.00313    3.35  0.00302
fire         0.04388    0.01032    4.25  0.00036
theft       -0.01722    0.00590   -2.92  0.00822
age          0.00938    0.00349    2.68  0.01390
income       0.06870    0.04216    1.63  0.11808

Residual standard error: 0.338 on 21 degrees of freedom
Multiple R-Squared: 0.791,       Adjusted R-squared: 0.741
F-statistic: 15.91 on 5 and 21 DF,  p-value: 1.594e-06
```

Any case with at least one missing value is omitted from the regression. Different statistical packages have different ways of handling missing observations. For example, the default behavior in S-plus would refuse to fit the model at all. You can see there are now only 21 degrees of freedom — almost half the data is lost. We can fill in the missing values by their variable means as in:

```
> cmeans <- apply(chmiss,2,mean,na.rm=T)
> cmeans
    race      fire     theft      age involact    income
35.60930 11.42444 32.65116 59.96905  0.64773 10.73587
> mchm <- chmiss
> for(i in c(1,2,3,4,6)) mchm[is.na(chmiss[,i]),i] <- cmeans[i]
```

We do not fill in missing values in the response because this is the variable we are trying to model. Now refit:

```
> g <- lm(involact ~ ., mchm)
> summary(g)
Coefficients:
            Value Std. Error t value Pr(>|t|)
(Intercept) 0.0707  0.5094    0.1387  0.8904
       race 0.0071  0.0027    2.6307  0.0122
       fire 0.0287  0.0094    3.0623  0.0040
      theft -0.0031 0.0027   -1.1139  0.2723
        age 0.0061  0.0032    1.8954  0.0657
     income -0.0271 0.0317   -0.8550  0.3979

Residual standard error: 0.3841 on 38 degrees of freedom
Multiple R-Squared: 0.6819,   Adjusted R-squared: 0.6401
F-statistic:  16.3 on 5 and 38 DF,  p-value: 1.409e-08
```

There are some important differences between these two fits. For example, `theft` and `age` are significant in the first fit, but not in the second. Also, the regression coefficients are now all closer to zero. The situation is analogous to the errors in variables case. The bias introduced by the fill-in method can be substantial and may not be compensated by the attendant reduction in variance.

We can also use regression methods to predict the missing values of the covariates. Let's try to fill in the missing race values:

```
> gr <- lm(race ~ fire+theft+age+income,chmiss)
> chmiss[is.na(chmiss$race),]
       race fire theft   age involact income
60646    NA  5.7    11  27.9      0.0 16.250
60651    NA 15.1    30  89.8      0.8 10.510
60616    NA 12.2    46  48.0      0.6  8.212
60617    NA 10.8    34  58.0      0.9 11.156
> predict(gr,chmiss[is.na(chmiss$race),])
   60646   60651   60616   60617
 -17.847  26.360  70.394  32.620
```

Notice that the first prediction is negative. Obviously, we need to put more work into the regression models used to fill in the missing values. One trick that can be applied when the response is bounded between zero and one is the logit transformation:

$$y \rightarrow \log(y/(1-y))$$

This transformation maps to the whole real line. We define the logit function and its inverse:

```
> logit <- function(x) log(x/(1-x))
> ilogit <- function(x) exp(x)/(1+exp(x))
```

We now fit the model with a logit-transformed response and then back transform the predicted values remembering to convert our percentages to proportions and vice versa at the appropriate times:

```
> gr <- lm(logit(race/100) ~ fire+theft+age+income,chmiss)
> ilogit(predict(gr,chmiss[is.na(chmiss$race),]))*100
   60646    60651    60616    60617
 0.41909 14.73202 84.26540 21.31213
```

We can see how our predicted values compare to the actual values:

```
> data(chredlin)
> chredlin$race[is.na(chmiss$race)]
[1]  1.0 13.4 62.3 36.4
```

So our first two predictions are good, but the other two are somewhat wide of the mark.

Like the mean fill-in method, the regression fill-in method will also introduce a bias towards zero in the coefficients while tending to reduce the variance. The success of the regression method depends somewhat on the collinearity of the predictors — the filled-in values will be more accurate the more collinear the predictors are.

For situations where there is a substantial proportion of missing data, I recommend that you investigate more sophisticated methods, likely using the EM algorithm or multiple imputation. The fill-in methods described in this chapter will be fine when only a few cases need to be filled, but will become less reliable as the proportion of missing cases increases. See Little and Rubin (2002) and Schafer (1997) for more.

CHAPTER 13

Analysis of Covariance

Predictors that are qualitative in nature, for example, eye color, are sometimes described as *categorical* or called *factors*. We wish to incorporate these predictors into the regression analysis. Analysis of covariance (ANCOVA) refers to regression problems where there is a mixture of quantitative and qualitative predictors.

Suppose we are interested in the effect of a medication on cholesterol level. We might have two groups; one receives the medication and the other, the default treatment. We could not treat this as a simple two-sample problem if we knew that the two groups differed with respect to age, as this would affect the cholesterol level. See Figure 13.1 for a simulated example. For the patients who received the medication, the mean reduction in cholesterol level was 0% while for those who did not, the mean reduction was 10%. So superficially it would seem that it would be better not to be treated. However, the treated group ranged in age from 50 to 70 while those who were not treated ranged in age between 30 and 50. We can see that once age is taken into account, the difference between treatment and control is again 10%, but this time in favor of the treatment.

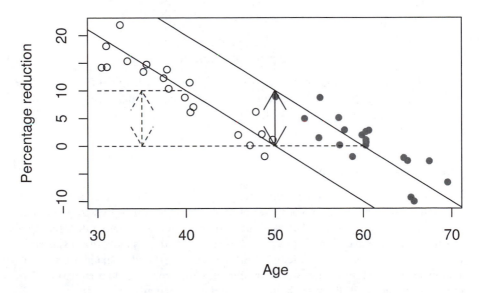

Figure 13.1 *Simulated example showing the confounding effect of a covariate. The patients who took the medication are marked with a solid dot while those who did not are marked with an empty dot.*

ANCOVA adjusts the groups for the age difference and then presents the effect of the medication. It can also be used when there are more than two groups and more than one covariate.

Our strategy is to incorporate the qualitative predictors within the $y = X\beta + \varepsilon$ framework. We can then use the estimation, inferential and diagnostic techniques that we have already learned. This avoids having to learn a different set of formulae for each new type of qualitative predictor configuration. To put qualitative predictors into the $y = X\beta + \varepsilon$ form we need to code the qualitative predictors. We consider a specific example where y is the change in cholesterol level, x is the age and:

$$d = \begin{cases} 0 & \text{did not take medication} \\ 1 & \text{took medication} \end{cases}$$

d is called a *dummy variable*. A variety of linear models may be considered here:

1. The same regression line for both groups: $y = \beta_0 + \beta_1 x + \varepsilon$ or is written in R as y ˜ x.

2. Separate regression lines for each group with the same slope: $y = \beta_0 + \beta_1 x + \beta_2 d + \varepsilon$ or is written in R as y ˜ x + d. In this case β_2 represents the vertical distance between the regression lines (i.e., the effect of the drug).

3. Separate regression lines for each group with the different slopes: $y = \beta_0 + \beta_1 x + \beta_2 d + \beta_3 x.d + \varepsilon$ or is written in R as y ˜ x + d + d:x or y ˜ x*d. To form the slope interaction term d:x in the X-matrix, multiply x by d elementwise. Any interpretation of the effect of the drug will now also depend on age.

Estimation and testing work just as they did before. Interpretation is much easier if we can eliminate the slope interaction term.

Other codings of d are possible, for instance:

$$d = \begin{cases} -1 & \text{did not take medication} \\ 1 & \text{took medication} \end{cases}$$

is used by some. This coding enables β_2 and β_3 to be viewed as differences from a response averaged over the two groups. Any other coding that assigned a different number to the two groups would also work, but interpretation of the estimated parameters would be more difficult.

13.1 A Two-Level Example

The data for this example come from a study of the effects of childhood sexual abuse on adult females reported in Rodriguez et al. (1997): 45 women treated at a clinic, who reported childhood sexual abuse (csa), were measured for post-traumatic stress disorder (ptsd) and childhood physical abuse (cpa) both on standardized scales. 31 women treated at the same clinic, who did not report childhood sexual abuse, were also measured. The full study was more complex than reported here and so readers interested in the subject matter should refer to the original article.

We take a look at the data and produce a summary subsetted by csa:

```
> data(sexab)
```

```
> sexab
        cpa       ptsd        csa
1   2.04786   9.71365      Abused
2   0.83895   6.16933      Abused
.....
75  2.85253   6.84304   NotAbused
76  0.81138   7.12918   NotAbused
> by(sexab,sexab$csa,summary)
sexab$csa: Abused
       cpa                ptsd                csa
 Min.   :-1.11    Min.   : 5.98     Abused   :45
 1st Qu.: 1.41    1st Qu.: 9.37     NotAbused: 0
 Median : 2.63    Median :11.31
 Mean   : 3.08    Mean   :11.94
 3rd Qu.: 4.32    3rd Qu.:14.90
 Max.   : 8.65    Max.   :18.99
---------------------------------------------------
sexab$csa: NotAbused
       cpa                ptsd                csa
 Min.   :-3.12    Min.   :-3.35     Abused   : 0
 1st Qu.:-0.23    1st Qu.: 3.54     NotAbused:31
 Median : 1.32    Median : 5.79
 Mean   : 1.31    Mean   : 4.70
 3rd Qu.: 2.83    3rd Qu.: 6.84
 Max.   : 5.05    Max.   :10.91
```

Now plot the data — see Figure 13.2:

```
> plot(ptsd ~ csa, sexab)
> plot(ptsd ~ cpa, pch=as.character(csa), sexab)
```

We see that those in the abused group have higher levels of PTSD than those in the nonabused in the left panel of Figure 13.2. We can test this difference:

```
> t.test(sexab$ptsd[1:45],sexab$ptsd[46:76])

        Welch Two Sample t-test

data:  sexab$ptsd[1:45] and sexab$ptsd[46:76]
t = 8.9006, df = 63.675, p-value = 8.803e-13
alt. hypothesis: true difference in means is not equal to 0
95 percent confidence interval:
 5.6189 8.8716
sample estimates:
mean of x mean of y
  11.9411    4.6959
```

and find that it is clearly significant. However, in the right panel of Figure 13.2 we see that there is positive correlation between PTSD and childhood *physical* abuse and in the numerical summary we see that those in the abused group suffered higher levels (3.08 vs. 1.31) of cpa than those in the nonabused group. This suggests physical abuse as an alternative explanation of higher PTSD in the sexually abused group.

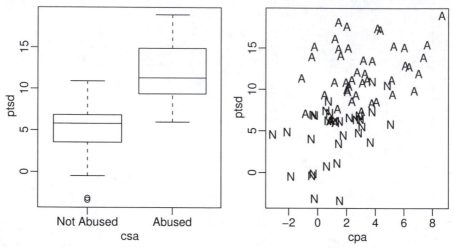

Figure 13.2 *PTSD comparison of abused and nonabused subjects on the left. A = Abused and N = NotAbused on the right.*

ANCOVA allows us to disentangle these two competing explanations. We fit the separate regression lines model. `ptsd ~ cpa*csa` is an equivalent model formula:

```
> g <- lm(ptsd ~ cpa+csa+cpa:csa,sexab)
> summary(g)
Coefficients:
                   Estimate Std. Error t value Pr(>|t|)
(Intercept)          10.557      0.806   13.09  < 2e-16
cpa                   0.450      0.208    2.16    0.034
csaNotAbused         -6.861      1.075   -6.38  1.5e-08
cpa:csaNotAbused      0.314      0.368    0.85    0.397

Residual standard error: 3.28 on 72 degrees of freedom
Multiple R-Squared: 0.583,        Adjusted R-squared: 0.565
F-statistic: 33.5 on 3 and 72 DF,  p-value: 1.13e-13
```

Because `csa` is nonnumeric, R automatically treats it as a qualitative variable and sets up a coding. We can discover the coding by examining the X-matrix:

```
> model.matrix(g)
     (Intercept)      cpa csaNotAbused cpa:csaNotAbused
1              1  2.04786            0          0.00000
2              1  0.83895            0          0.00000
......
75             1  2.85253            1          2.85253
76             1  0.81138            1          0.81138
```

We see that "Abused" is coded as zero and "NotAbused" is coded as one. The default choice is made alphabetically. This means that "Abused" is the *reference level* here and that the parameters represent the difference between "NotAbused" and this reference level. In this case, it would be slightly more convenient if the coding was reversed. The interaction term `cpa:csaNotAbused` is represented in the fourth column of the matrix as the product of the second and third columns which represents the terms from which the interaction is formed.

We see that the model can be simplified because the interaction term is not significant. We reduce to this model:

```
> g <- lm(ptsd ~ cpa+csa,sexab)
> summary(g)
Coefficients:
                Estimate Std. Error t value Pr(>|t|)
(Intercept)       10.248       0.719   14.26  < 2e-16
cpa                0.551       0.172    3.21    0.002
csaNotAbused      -6.273       0.822   -7.63  6.9e-11

Residual standard error: 3.27 on 73 degrees of freedom
Multiple R-Squared: 0.579,        Adjusted R-squared: 0.567
F-statistic: 50.1 on 2 and 73 DF,  p-value: 2e-14
```

No further simplification is possible because the remaining predictors are statistically significant.

Put the two parallel regression lines on the plot, as seen in the left panel of Figure 13.3.

```
> plot(ptsd ~ cpa, pch=as.character(csa), sexab)
> abline(10.248,0.551)
> abline(10.248-6.273,0.551,lty=2)
```

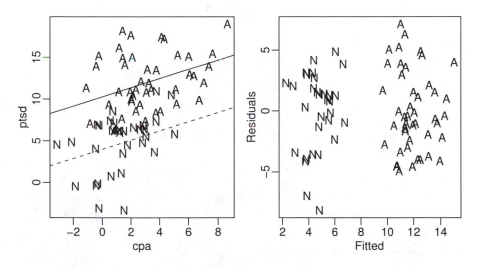

Figure 13.3 *Model fit shown on the left and fitted vs. residuals plot on the right. A = Abused and N = NotAbused.*

The slope of both lines is 0.551, but the "Abused" line is 6.273 higher than the "NonAbused." From the t-test above, the estimated effect of childhood sexual abuse is $11.9411 - 4.6959 = 7.2452$. So after adjusting for the effect of childhood physical abuse, our estimate of the effect of childhood sexual abuse on PTSD is mildly reduced.

We can also compare confidence intervals for the effect of `csa`:

```
> confint(g)[3,]
   2.5 %   97.5 %
-7.9108  -4.6347
```

compared to the $(5.6189, 8.8716)$ found for the unadjusted difference. In this particular case, the confidence intervals are about the same width. In other cases, particularly designed experiments, adjusting for a covariate can increase the precision of the estimate of an effect.

The usual diagnostics should be checked. It is worth checking whether there is some difference related to the categorical variable as we do here:

```
> plot(fitted(g),residuals(g),pch=as.character(sexab$csa),
  xlab="Fitted",ylab="Residuals")
```

We see in the right panel of Figure 13.3 that there are no signs of heteroscedasticity. Furthermore, because the two groups happen to separate, we can also see that the variation in the two groups is about the same. If this were not so, we would need to make some adjustments to the analysis, possibly using weights.

For convenience, you can change the reference level:

```
> sexab$csa <- relevel(sexab$csa,ref="NotAbused")
> g <- lm(ptsd ~ cpa+csa,sexab)
> summary(g)
Coefficients:
             Estimate Std. Error t value Pr(>|t|)
(Intercept)     3.975      0.629    6.32  1.9e-08
cpa             0.551      0.172    3.21    0.002
csaAbused       6.273      0.822    7.63  6.9e-11

Residual standard error: 3.27 on 73 degrees of freedom
Multiple R-Squared: 0.579,        Adjusted R-squared: 0.567
F-statistic: 50.1 on 2 and 73 DF,  p-value: 2e-14
```

Although some of the coefficients have different numerical values, this coding leads to the same conclusion as before.

Finally, we should point out that childhood physical abuse might not be the only factor that is relevent to assessing the effects of childhood sexual abuse. It is quite possible that the two groups differ according to other variables such as socioeconomic status and age. Issues such as these were addressed in Rodriguez et al. (1997).

13.2 Coding Qualitative Predictors

A more extensive use of dummy variables is needed for factors with more than two levels. Let B be an $n \times k$ dummy variable matrix where $B_{ij} = 1$ if case i falls in class j and is zero otherwise. We might use B to form part of the model matrix. However, the row sums of B are all one. Since an intercept term would also be represented by a column of ones, all the parameters would not be identifiable.

Removing the intercept term is one solution, but this will not work well if there is more than one factor. A more general solution is to reduce the rank of the dummy variable matrix. Simply deleting one column would do, but any solution that avoids collinear dummy variables will work. The choice should be based on convenience of interpretation and numerical stability.

The coding is determined by a *contrast matrix* C which has dimension $k \times (k-1)$. Some examples of C are given below. The contribution to the model matrix is then given by BC. Other columns of the model matrix might include a column of ones for the intercept and perhaps other predictors.

Treatment coding

Consider a four-level factor that will be coded using three dummy variables. This contrast matrix describes the coding, where the columns represent the dummy variables and the rows represent the levels:

```
> contr.treatment(4)
  2 3 4
1 0 0 0
2 1 0 0
3 0 1 0
4 0 0 1
```

This treats level one as the standard level to which all other levels are compared so a control group, if one exists, would be appropriate for this level. The parameter for the dummy variable then represents the difference between the given level and the first level. R assigns levels to a factor in alphabetical order by default. Treatment coding is the default choice for R.

Helmert coding

The coding here is:

```
> contr.helmert(4)
   [,1] [,2] [,3]
1   -1   -1   -1
2    1   -1   -1
3    0    2   -1
4    0    0    3
```

If there are equal numbers of observations in each level (a balanced design) then the dummy variables will be orthogonal to each other and to the intercept. This coding is not as nice for interpretation except in some special cases. It is the default choice in S-plus.

There are other choices of coding — anything that spans the $k-1$ dimensional space will work. The choice of coding does not affect the R^2, $\hat{\sigma}^2$ and overall F-statistic. It does affect the $\hat{\beta}$ and you do need to know what the coding is before making conclusions about $\hat{\beta}$.

When there is an interaction between two terms represented by model matrix components X_1 and X_2, we compute the element-wise product of every column of X_1 with every column of X_2 to get the representation of the interaction in the model matrix.

13.3 A Multilevel Factor Example

The data for this example come from a study on the sexual activity and the life span of male fruitflies by Partridge and Farquhar (1981): 125 fruitflies were divided randomly into five groups of 25 each. The response was the longevity of the fruitfly in days. One group was kept solitary, while another was kept individually with a virgin female each day. Another group was given eight virgin females per day. As an additional control, the fourth and fifth groups were kept with one or eight pregnant females per day. Pregnant fruitflies will not mate. The thorax length of each male was measured as this was known to affect longevity. The five groups are labeled isolated, low, high, one and many respectively. The purpose of the analysis is to determine the difference between the five groups if any. We start with a plot of the data, as seen in Figure 13.4.

```
> data(fruitfly)
> plot(longevity ~ thorax, fruitfly, pch=unclass(activity))
> legend(0.63,100,levels(fruitfly$activity),pch=1:5)
```

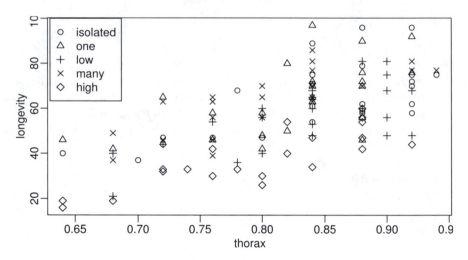

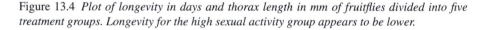

Figure 13.4 *Plot of longevity in days and thorax length in mm of fruitflies divided into five treatment groups. Longevity for the high sexual activity group appears to be lower.*

We fit and summarize the most general linear model:

```
> g <- lm(longevity ~ thorax*activity, fruitfly)
> summary(g)
Coefficients:
```

	Estimate	Std. Error	t value	Pr(>|t|)
(Intercept)	-50.242	21.801	-2.30	0.023
thorax	136.127	25.952	5.25	7.3e-07
activityone	6.517	33.871	0.19	0.848
activitylow	-7.750	33.969	-0.23	0.820
activitymany	-1.139	32.530	-0.04	0.972

```
activityhigh              -11.038      31.287    -0.35     0.725
thorax:activityone         -4.677      40.652    -0.12     0.909
thorax:activitylow          0.874      40.425     0.02     0.983
thorax:activitymany         6.548      39.360     0.17     0.868
thorax:activityhigh       -11.127      38.120    -0.29     0.771
```

```
Residual standard error: 10.7 on 114 degrees of freedom
Multiple R-Squared: 0.653,          Adjusted R-squared: 0.626
F-statistic: 23.9 on 9 and 114 DF,  p-value: <2e-16
```

Since "isolated" is the reference level, the fitted regression line within this group is `longevity=-50.2+136.1*thorax`. For "many," it is `longevity= (-50.2+ 6.5) + (136.1-4.7)*thorax`. Similar calculations can be made for the other groups. Examine:

```
> model.matrix(g)
```

to see how the coding is done. Some diagnostics should be examined by:

```
> plot(g)
```

There is perhaps some mild heteroscedasticity, but we will let this be until later for ease of presentation. Now see whether the model can be simplified:

```
> anova(g)
Analysis of Variance Table
```

```
Response: longevity
                Df Sum Sq Mean Sq F value  Pr(>F)
thorax           1  15003   15003 130.73 < 2e-16
activity         4   9635    2409  20.99 5.5e-13
thorax:activity  4     24       6   0.05       1
Residuals      114  13083     115
```

This is a sequential analysis of variance (ANOVA) table. Starting from a null model, terms are added and sequentially tested. The models representing the null and alternatives are listed in Table 13.1. In ANCOVA, we wish to successively simplify the

Null	Alternative
y~1	y~thorax
y~thorax	y~thorax+activity
y~thorax+activity	y~thorax+activity+thorax:activity

Table 13.1 *Models compared in the sequential ANOVA.*

full model and then interpret the result. The interaction term `thorax:activity` is not significant indicating that we can fit the same slope within each group. No further simplification is possible.

We notice that the F-statistic for the test of the interaction term is very small and so the p-value is not exactly one (due to the rounding), but is very close:

```
> 1-pf(0.05,4,114)
[1] 0.99525
```

For these data, the fitted regression lines to the five groups happen to be very close to parallel. This can, of course, just happen by chance. In some other cases, unusually large p-values have been used as evidence that data have been tampered with or "cleaned" to improve the fit. Most famously, Ronald Fisher suspected Gregor Mendel of fixing the data in some genetics experiments because the data seemed too good to be true. See Fisher (1936).

We now refit without the interaction term:

```
> gb <- lm(longevity ~ thorax+activity, fruitfly)
```

Do we need both `thorax` and `activity`? We could use the output above which suggests both terms are significant. However, `thorax` is tested by itself and then `activity` is tested once `thorax` is entered into the model. We might prefer to check whether each predictor is significant once the other has been taken into account. We can do this using:

```
> drop1(gb,test="F")
Single term deletions

Model:
longevity ~ thorax + activity
          Df Sum of Sq   RSS   AIC F value    Pr(F)
<none>                  13107   590
thorax     1     12368 25476   670  111.3 < 2e-16
activity   4      9635 22742   650   21.7 2.0e-13
```

The `drop1()` command tests each term relative to the full model. This shows that both terms are significant even after allowing for the effect of the other. Now examine the model coefficients:

```
> summary(gb)

Coefficients:
             Estimate Std. Error t value Pr(>|t|)
(Intercept)    -48.75      10.85   -4.49  1.6e-05
thorax         134.34      12.73   10.55  < 2e-16
activityone      2.64       2.98    0.88    0.379
activitylow     -7.01       2.98   -2.35    0.020
activitymany     4.14       3.03    1.37    0.174
activityhigh   -20.00       3.02   -6.63  1.0e-09

Residual standard error: 10.5 on 118 degrees of freedom
Multiple R-Squared: 0.653,     Adjusted R-squared: 0.638
F-statistic: 44.4 on 5 and 118 DF,  p-value: <2e-16
```

"Isolated is the reference level. We see that the intercepts of "one" and "many" are not significantly different from many. We also see that the low sexual activity group, "low," survives about seven days less. The p-value is 0.02 and is enough for statistical significance if only one comparison is made. However, we are making more than one comparison, and so, as with outliers, a Bonferroni-type adjustment might be considered. This would erase the statistical significance of the difference.

However, the high sexual activity group, "high," has a life span 20 days less than the reference group and this is strongly significant.

Returning to the diagnostics:

```
> plot(residuals(gb) ~ fitted(gb),
  pch=unclass(fruitfly$activity))
```

is seen in the first panel of Figure 13.5. We have some nonconstant variance although it does not appear to be related to the five groups. A log transformation can remove the heteroscedasticity:

```
> gt <- lm(log(longevity) ~ thorax+activity, fruitfly)
> plot(residuals(gt) ~ fitted(gt),
  pch=unclass(fruitfly$activity))
```

as seen in the second panel of Figure 13.5. One disadvantage of transformation is

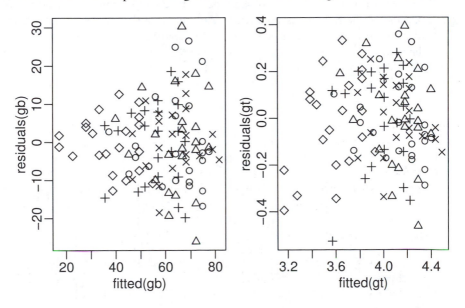

Figure 13.5 *Diagnostic plots for the fruitfly data before and after log transformation of the response.*

that it can make interpretation of the model more difficult. Let's examine the model fit:

```
> summary(gt)
Coefficients:
              Estimate Std. Error t value Pr(>|t|)
(Intercept)     1.8442     0.1988    9.28  1.0e-15
thorax          2.7215     0.2333   11.67  < 2e-16
activityone     0.0517     0.0547    0.95    0.346
activitylow    -0.1239     0.0546   -2.27    0.025
activitymany    0.0879     0.0555    1.59    0.116
activityhigh   -0.4193     0.0553   -7.59  8.4e-12
```

```
Residual standard error: 0.193 on 118 degrees of freedom
Multiple R-Squared: 0.702,        Adjusted R-squared: 0.69
F-statistic: 55.7 on 5 and 118 DF,   p-value: <2e-16
```

Notice that the R^2 is higher for this model, but the p-values are similar. Because of the log transformation, we can interpret the coefficients as having a multiplicative effect:

```
> exp(coef(gt)[3:6])
 activityone   activitylow activitymany activityhigh
    1.05311       0.88350      1.09189      0.65754
```

Compared to the reference level, we see that the high sexual activity group has 0.66 times the life span (i.e, 34% less).

Why did we include `thorax` in the model? Its effect on longevity was known, but because of the random assignment of the flies to the groups, this variable will not bias the estimates of the effects of the activities. We can verify that `thorax` is unrelated to the activities:

```
> gh <- lm(thorax ~ activity, fruitfly)
> anova(gh)
Analysis of Variance Table

Response: thorax
           Df Sum Sq Mean Sq F value Pr(>F)
activity    4  0.026   0.006    1.11   0.36
Residuals 119  0.685   0.006
```

However, look what happens if we omit `thorax` from the model for `longevity`:

```
> gu <- lm(log(longevity) ~ activity, fruitfly)
> summary(gu)
Coefficients:
              Estimate Std. Error t value Pr(>|t|)
(Intercept)     4.1193     0.0564   72.99  < 2e-16
activityone     0.0234     0.0798    0.29     0.77
activitylow    -0.1195     0.0798   -1.50     0.14
activitymany    0.0240     0.0806    0.30     0.77
activityhigh   -0.5172     0.0798   -6.48  2.2e-09

Residual standard error: 0.282 on 119 degrees of freedom
Multiple R-Squared: 0.359,        Adjusted R-squared: 0.338
F-statistic: 16.7 on 4 and 119 DF,   p-value: 6.96e-11
```

The magnitude of the effects do not change that much but the standard errors are substantially larger. The value of including `thorax` in this model is to increase the precision of the estimates.

Exercises

1. Using the `teengamb` data, model `gamble` as the response and the other variables as predictors. Take care to investigate the possibility of interactions between `sex` and the other predictors. Interpret your final model.

2. Using the `infmort` data, find a simple model for the infant mortality in terms of the other variables. Be alert for transformations and unusual points. Interpret your model by explaining what the regression parameter estimates mean.

3. Plot the `ToothGrowth` data with `len` as the response. Fit a linear model to the data and check for possible simplification. Display the fitted regression lines on top of the data.

4. Investigate whether the `side` variable can be used in a model for all the `chredlin` data rather than as a subsetting variable as in the analysis in the text.

5. Find a good model for the `uswages` data with `wages` as the response and all the other variables as predictors.

CHAPTER 14

One-Way Analysis of Variance

In an analysis of variance (ANOVA), the idea is to partition the overall variance in the response into that due to each of the factors and the error. The traditional approach is to directly estimate these components. However, we take a regression-based approach by putting the model into the $y = X\beta + \varepsilon$ and then using the inferential methods we have learned earlier in this book.

The terminology used in ANOVA-type problems is sometimes different. Predictors are now all qualitative and are now typically called *factors*, which have some number of *levels*. The regression parameters are now often called *effects*. We shall consider only models where the parameters are considered fixed, but unknown — called *fixed-effects* models. *Random-effects* models are used where parameters are taken to be random variables and are not covered in this text.

14.1 The Model

Suppose we have a factor α occurring at $i = 1, \ldots, I$ levels, with $j = 1, \ldots, J_i$ observations per level. We use the model:

$$y_{ij} = \mu + \alpha_i + \varepsilon_{ij}$$

Not all the parameters are identifiable. For example, we could add some constant to μ and subtract the same constant from each α_i and the model would be unchanged. Some restriction is necessary. Here are some possibilities:

1. Set $\mu = 0$ and then use I different dummy variables to estimate α_i for $i = 1, \ldots, I$.

2. Set $\alpha_1 = 0$, then μ represents the expected mean response for the first level and α_i for $i \neq 1$ represents the difference between level i and level one. Level one is then called the *reference level* or *baseline level*. This corresponds to the use of treatment contrasts as discussed in the previous chapter.

3. Set $\sum_i \alpha_i = 0$, now μ represents the mean response over all levels and α_i, the difference from that mean. Because $\alpha_I = -\sum_{i=1}^{I-1} \alpha_i$, we do not need to estimate α_I directly because it can be determined from the other estimates. This approach requires the use of sum contrasts.

Some preliminary graphical analysis is appropriate before fitting. A side-by-side boxplot is often recommended, although strip plots are better for smaller datasets. Look for equality of variance, the need for transformation of the response and outliers. It is not worth considering diagnostics for influence, as the leverages depend explicitly on J_i. If there is only one observation for a given level, that is $J_i = 1$, then the estimate of the effect for that level will be based on that single point. That point is clearly influential without further investigation.

The choice of constraint from those listed above or otherwise will determine the coding used to generate the X-matrix. Once that is done, the parameters (effects) can be estimated in the usual way along with standard errors. No matter which constraint and coding choice is made, the fitted values and residuals will be the same.

Once the effects are estimated, the natural first step is to test for differences in the levels of the factor. An explicit statement of the null and alternative hypotheses would depend on the coding used. If we use the treatment coding with a reference level, then the null hypothesis would require that $\alpha_2 = \cdots = \alpha_I = 0$. For other codings, the statement would differ. It is simpler to state the hypotheses in terms of models:

$$H_0: \quad y_{ij} \quad = \quad \mu + \varepsilon_{ij}$$
$$H_1: \quad y_{ij} \quad = \quad \mu + \alpha_i + \varepsilon_{ij}$$

We compute the residual sum of squares and degrees of freedom for the two models and then use the same F-test as we have used for regression. The outcome of this test will be the same no matter what coding/restriction we use. If we do not reject the null, we are almost done — we must still check for a possible transformation of the response and outliers. If we reject the null, we must investigate which levels differ.

14.2 An Example

Our example dataset comes from a study of blood coagulation times: 24 animals were randomly assigned to four different diets and the samples were taken in a random order. These data come from Box, Hunter, and Hunter (1978):

```
> data(coagulation)
> coagulation
   coag diet
1    62    A
2    60    A
...etc...
23   63    D
24   59    D
```

The first step is to plot the data. We compare boxplots and strip plots:

```
> plot(coag ~ diet, coagulation, ylab="coagulation time")
> with(coagulation, stripchart(coag ~ diet, vertical=TRUE,
    method="stack",xlab="diet",ylab="coagulation time"))
```

See Figure 14.1. The strip plot is preferred here because the boxplot displays poorly when there are few datapoints. For larger datasets, the boxplot might be better. We see no outliers, skewness or unequal variance. Some judgment is required because when there are very few datapoints, even when the variances truly are equal in the groups, we can expect some noticeable variability. In this case, there are no obvious problems.

Now let's fit the model using the default treatment coding:

```
> g <- lm(coag ~ diet, coagulation)
```

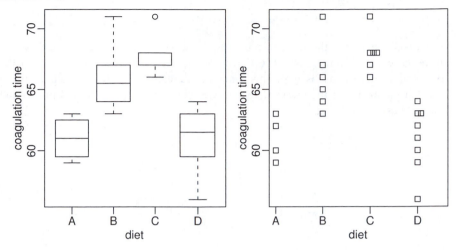

Figure 14.1 *A boxplot and a strip plot of blood coagulation data.*

```
> summary(g)
Coefficients:
              Estimate Std. Error  t value Pr(>|t|)
(Intercept)   6.10e+01   1.18e+00    51.55  < 2e-16
dietB         5.00e+00   1.53e+00     3.27  0.00380
dietC         7.00e+00   1.53e+00     4.58  0.00018
dietD        -1.07e-14   1.45e+00  -7.4e-15  1.00000

Residual standard error: 2.37 on 20 degrees of freedom
Multiple R-Squared: 0.671,        Adjusted R-squared: 0.621
F-statistic: 13.6 on 3 and 20 DF,  p-value: 4.66e-05
```

We conclude from the small p-value for the F-statistic that there is some difference between the groups. Group A is the reference level and has a mean of 61, groups B, C and D are 5, 7 and 0 seconds larger, respectively, on average. Examine the design matrix to understand the coding:

```
> model.matrix(g)
```

We can fit the model without an intercept term as in:

```
> gi <- lm(coag ~ diet -1, coagulation)
> summary(gi)
Coefficients:
       Estimate Std. Error t value Pr(>|t|)
dietA    61.000      1.183    51.5   <2e-16
dietB    66.000      0.966    68.3   <2e-16
dietC    68.000      0.966    70.4   <2e-16
dietD    61.000      0.837    72.9   <2e-16

Residual standard error: 2.37 on 20 degrees of freedom
Multiple R-Squared: 0.999,        Adjusted R-squared: 0.999
```

```
F-statistic: 4.4e+03 on 4 and 20 DF,     p-value:    0
```

We can directly read the level means. The R^2 is not correctly calculated because of the absence of an intercept. The F-test corresponds to a null hypothesis that the expected mean response is zero. This is not an interesting test. To generate the usual test that the means of the levels are equal, we would need to fit the null model and compare using an F-test:

```
> gnull <- lm(coag ~ 1, coagulation)
> anova(gnull,gi)
Analysis of Variance Table

Model 1: coag ~ 1
Model 2: coag ~ diet - 1
  Res.Df RSS Df Sum of Sq    F   Pr(>F)
1     23 340
2     20 112  3       228 13.6 4.7e-05
```

We get the same F-statistic and p-value as in the first coding.

We can also use a sum coding:

```
> options(contrasts=c("contr.sum","contr.poly"))
> gs <- lm(coag ~ diet , coagulation)
> summary(gs)

Coefficients:
             Estimate Std. Error t value Pr(>|t|)
(Intercept)    64.000      0.498  128.54  < 2e-16
diet1          -3.000      0.974   -3.08  0.00589
diet2           2.000      0.845    2.37  0.02819
diet3           4.000      0.845    4.73  0.00013

Residual standard error: 2.37 on 20 degrees of freedom
Multiple R-Squared: 0.671,      Adjusted R-squared: 0.621
F-statistic: 13.6 on 3 and 20 DF,  p-value: 4.66e-05
```

So the estimated overall mean response is 64 while the estimated mean response for A is three less than the overall mean, that is 61. Similarly, the means for B and C are 66 and 68, respectively. Since we are using the sum constraint, we compute $\hat{\alpha}_D = -(-3+2+4) = -3$ so the mean for D is $64-3 = 61$. Notice that $\hat{\sigma}$ and the F-statistic are the same as before.

So we can use any of these three methods and obtain essentially the same results. The constraint $\mu = 0$ is least convenient since an extra step is needed to generate the F-test. Furthermore, the approach would not extend well to experiments with more than one factor, as additional constraints would be needed. The other two methods can be used according to taste. The treatment coding is most appropriate when the reference level is set to a possible control group. I will use the treatment coding for the rest of this book.

14.3 Diagnostics

There are fewer diagnostics to do for ANOVA models, but it is still important to plot the residuals and fitted values and to make the Q–Q plot of the residuals. It makes no sense to transform the predictor, but it is reasonable to consider transforming the response. Diagnostics are shown in Figure 14.2:

```
> qqnorm(residuals(g))
> plot(jitter(fitted(g)),residuals(g),
  xlab="Fitted",ylab="Residuals")
```

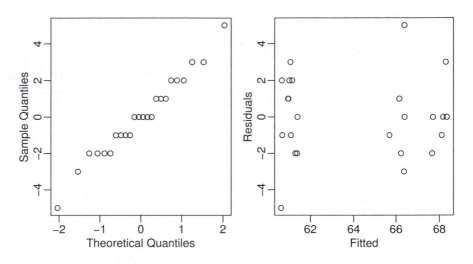

Figure 14.2 *Diagnostics for the blood coagulation model.*

Because the data are integers and the fitted values turn out to be integers also, some discreteness is obvious in the Q–Q plot. Of course, discrete data cannot be normally distributed. However, here it is approximately normal and so we can go ahead with the inference without much concern. The discreteness in the residuals and fitted values shows up in the residual-fitted plot. We have jittered the points so that they can seen separately.

The assumption of homogeneity of the error variance can be examined using Levene's test. It computes the absolute values of the residuals and uses these as the response in a new one-way ANOVA. To reduce the possible influence of outliers, group medians rather than means should be used. A significant difference would indicate nonconstant variance.

There are other tests, but this one is quite insensitive to nonnormality and is simple to execute. Most tests and confidence intervals (CIs) are relatively insensitive to nonconstant variance so there is no need to take action unless the Levene test is significant at the 1% level.

Applying this to the diet data, we find:

```
> med <- with(coagulation,tapply(coag,diet,median))
```

```
> ar <- with(coagulation,abs(coag -med[diet]))
> anova(lm(ar ~ diet,coagulation))
Analysis of Variance Table

Response: ar
          Df Sum Sq Mean Sq F value Pr(>F)
diet       3    4.3     1.4    0.65   0.59
Residuals 20   44.5     2.2
```

Since the p-value is large, we conclude that there is no evidence of a nonconstant variance.

14.4 Pairwise Comparisons

After detecting some difference in the levels of the factor, interest centers on which levels or combinations of levels are different. It does not make sense to ask whether a particular level is significant since this begs the question of "significantly different from what?" Any meaningful test must involve a comparison of some kind.

A pairwise comparison of level i and j can be made using a CI for $\alpha_i - \alpha_j$ using:

$$\hat{\alpha}_i - \hat{\alpha}_j \pm t_{df}^{\alpha/2} se(\hat{\alpha}_i - \hat{\alpha}_j)$$

where $se(\hat{\alpha}_i - \hat{\alpha}_j) = \hat{\sigma}\sqrt{1/J_i + 1/J_j}$ and $df = n - I$ in this case. A test for $\alpha_i = \alpha_j$ amounts to seeing whether zero lies in this interval or not. For example, a 95% CI for the $B - A$ difference above is:

```
> qt(0.975,20)
[1] 2.0860
> c(5-2.086*1.53,5+2.086*1.53)
[1] 1.8084 8.1916
```

Since zero is not in the interval, the difference is significant. This is fine for just one test, but we are likely to be interested in more than one comparison. Suppose we do all possible pairwise tests when $\alpha = 5\%$ and the null hypothesis is in fact true. In the blood coagulation data, there are four levels and so six possible pairwise comparisons. Even if there was no difference between the four levels, there is still about a 20% chance that at least one significant difference will be found.

For experiments with more levels, the true type I error gets even higher. Using the t-based CIs for multiple comparisons is called the least significant difference (LSD) method, but it can hardly be recommended. Now one might be tempted to argue that we could choose which comparisons are interesting and so reduce the amount of testing and thus the magnitude of the problem. If we only did a few tests, then the Bonferroni adjustment (see Section 4.2.2) could be used to make a simple correction. However, the determination of which comparisons are "interesting" is usually made after seeing the fitted model. This means that all other comparisons are implicitly made even if they are not explicitly computed. On the other hand, if it can be argued that the comparisons were decided before seeing the fit, then we could make the case for the simple adjustment. However, this is rarely the case and furthermore it might be difficult to convince others that this really was your intention. We must usually find a way to adjust for *all* pairwise comparisons.

There are many ways to make the adjustment, but *Tukey's honest significant difference (HSD)* is the easiest to understand. It depends on the studentized range distribution which arises as follows. Let $X_1, \ldots, X_n$ be i.i.d. $N(\mu, \sigma^2)$ and let $R = \max_i X_i - \min_i X_i$ be the range. Then $R/\hat{\sigma}$ has the studentized range distribution $q_{n,\nu}$ where ν is the number of degrees of freedom used in estimating σ.

The Tukey CIs are:

$$\hat{\alpha}_i - \hat{\alpha}_j \pm \frac{q_{l,df}}{\sqrt{2}} se(\hat{\alpha}_i - \hat{\alpha}_j)$$

When the sample sizes J_i are very unequal, Tukey's HSD test may become too conservative. We compute the Tukey HSD bands for the diet data. First, we need the critical value from the studentized range distribution:

```
> qtukey(0.95,4,20)/sqrt(2)
[1] 2.7989
```

and then the interval for the $B - A$ difference is:

```
> c(5-2.8*1.53,5+2.8*1.53)
[1] 0.716 9.284
```

A convenient way to obtain all the intervals is:

```
> TukeyHSD(aov(coag ~ diet, coagulation))
  Tukey multiple comparisons of means
    95% family-wise confidence level
```

```
$diet
          diff        lwr       upr
A-B -5.0000e+00  -9.2754  -0.72455
C-B  2.0000e+00  -1.8241   5.82407
D-B -5.0000e+00  -8.5771  -1.42291
C-A  7.0000e+00   2.7246  11.27545
D-A -1.4211e-14  -4.0560   4.05604
D-C -7.0000e+00 -10.5771  -3.42291
```

We find that only the $A - D$ and $B - C$ differences are not significant as the corresponding intervals contain zero. The Bonferroni-based bands would have been just slightly wider:

```
> qt(1-.05/12,20)
[1] 2.9271
```

We divide by 12 here because there are six possible pairwise differences and we want a two-sided CI.

The Tukey method assumes the worst by focusing on the largest difference. There are other competitors like the Newman–Keuls, Duncan's multiple range and the Waller–Duncan procedure, which are less pessimistic or do not consider all possible pairwise comparisons. For a detailed description of the many available alternatives see Hsu (1996). Some other pairwise comparison tests may be found in the R package multcomp.

A contrast among the effects $\alpha_1, \ldots, \alpha_I$ is a linear combination $\sum_i c_i \alpha_i$ where the c_i are known and $\sum_i c_i = 0$. For example:

1. $\alpha_1 - \alpha_2$ is a contrast with $c_1 = 1, c_2 = -1$ and the other $c_i = 0$. All pairwise differences are contrasts.

2. $\alpha_1 - (\alpha_2 + \alpha_3 + \alpha_4)/3$ with $c_1 = 1$ and $c_2 = c_3 = c_4 = -1/3$ and the other $c_i = 0$. This contrast might be interesting if level one were the control and two, three and four were alternative treatments.

The need for contrasts arises much less often than pairwise comparisons and the Scheffé method must be used.

Exercises

1. Using the `pulp` data, determine whether there are any differences between the operators. What is the nature of these differences?

2. Determine whether there are differences in the weights of chickens according to their feed in the `chickwts` data. Perform all necessary model diagnostics.

3. Using the `PlantGrowth` data, determine whether there are any differences between the groups. What is the nature of these differences? Test for a difference between the average of the two treatments and the control.

4. Using the `infmort` data, perform a one-way ANOVA with `income` as the response and `region` as the predictor. Which pairs of regions are different? Now check for a good transformation on the response and repeat the comparison.

CHAPTER 15

Factorial Designs

A factorial design has some number of factors occurring at some number of levels. In a full factorial design, all possible combinations of the levels of the factors occur at least once. We start with experiments involving two factors and then look at designs with more than two factors.

15.1 Two-Way ANOVA

Suppose we have two factors, α at I levels and β at J levels. Let n_{ij} be the number of observations at level i of α and level j of β and let those observations be $y_{ij1}, y_{ij2}, \cdots$ etc. A complete layout has $n_{ij} \geq 1$ for all i, j. A balanced layout requires that $n_{ij} = n$.

The most general model that may be considered is:

$$y_{ijk} = \mu + \alpha_i + \beta_j + (\alpha\beta)_{ij} + \varepsilon_{ijk}$$

As in the one-way layout, not all the parameters are identifiable. The interaction effect $(\alpha\beta)_{ij}$ is interpreted as that part of the mean response not attributable to the additive effect of α_i and β_j. For example, you may enjoy strawberries and cream individually, but the combination is far superior. In contrast, you may like fish and ice cream, but not together.

If the main effects α and β generate design matrices X_α and X_β, then the design matrix for the $\alpha\beta$ interaction is given by collecting the element-wise products of all columns of X_α with all columns of X_β to form a matrix with $(I-1)(J-1)$ columns.

A significant interaction makes the model hard to interpret, as $\hat{\alpha}$ cannot be studied independent of $\hat{\beta}$. A comparison of the levels of α will depend on the level of β. Consider the following two layouts of $\hat{\mu}_{ij}$ in a 2×2 example:

	Male	Female		Male	Female
Drug 1	3	5		2	1
Drug 2	1	2		1	2

The response is a measure of performance. In the case on the left, we can say that drug 1 is better than drug 2 although the interaction means that its superiority over drug 2 depends on the gender. In the case on the right, which drug is best depends on the gender. We plot this in Figure 15.1. We see that in both cases the lines are not parallel, indicating interaction. The superiority of drug 1 is clear in the first plot and an ambiguous conclusion is seen in the second. Make plots like this when you want to understand an interaction effect.

When the interaction is significant, the main effects cannot be defined in an obvious and universal way. For example, we could define the gender effect as the effect

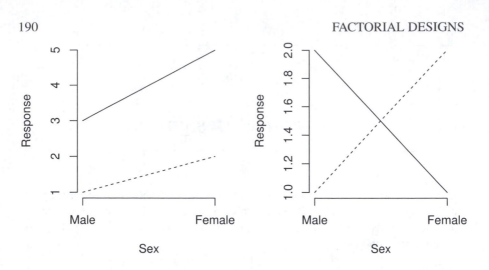

Figure 15.1 *Two 2×2 tables with the response plotted by the factors, sex on the horizontal axis and drug 1 as the solid line and drug 2 as the dotted line.*

for females, the effect for males, the effect for the average of males and females or something else. If there was no interaction effect, the gender effect could be defined unambiguously.

When you have a significant interaction, an interaction plot of the fitted values can make it easier to describe the nature and impact of the effect. Alternatively, you can fit a model:

$$y_{ijk} = \mu_{ijk} + \varepsilon_{ijk}$$

and then treat the data as a one-way ANOVA with *IJ* levels. Obviously this makes for more complex comparisons, but this is unavoidable when interactions exist.

15.2 Two-Way ANOVA with One Observation per Cell

Mazumdar and Hoa (1995) report an experiment to test the strength of a thermoplastic composite depending on the power of a laser and the speed of a tape:

```
> data(composite)
> summary(composite)
     strength        laser          tape
 Min.   :20.6    40W:3     slow   :3
 1st Qu.:28.0    50W:3     medium :3
 Median :29.8    60W:3     fast   :3
 Mean   :31.0
 3rd Qu.:35.7
 Max.   :39.6
```

We can fit a model with just the main effects as:

```
> g <- lm(strength ~ laser + tape, composite)
```

However, if we tried to add an interaction term, we would have as many observations as parameters. The parameters could be estimated, but no further inference would be

possible. Nevertheless, the possibility of an important interaction exists. There are
two ways we can check for this.

The interaction can be checked graphically using an *interaction plot*. We plot the
cell means on the vertical axis and the factor α on the horizontal. We join the points
with same level of β. The role of α and β can be reversed. Parallel lines on the plot
are a sign of a lack of interaction as seen in the plots in Figure 15.2:

```
> with(composite,interaction.plot(laser,tape,strength,legend=F))
> with(composite,interaction.plot(tape,laser,strength,legend=F))
```

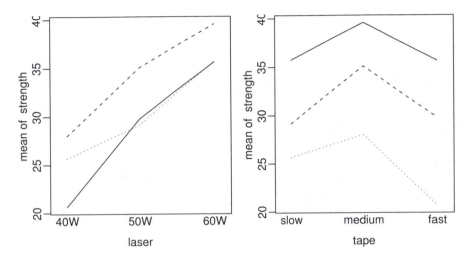

Figure 15.2 *Interaction plots for the composite data.*

Because of random variation, we cannot expect the lines to be exactly parallel. Allowing for a certain amount of noise, these lines are approximately parallel, so we
conclude there is no substantial interaction between the factors.

Tukey's nonadditivity test provides another way of investigating an interaction.
The model:

$$y_{ij} = \mu + \alpha_i + \beta_j + \phi\alpha_i\beta_j + \varepsilon_{ijk}$$

is fit to the data and then we test if $\phi = 0$. This is a nonlinear model because it
involves the product of parameters, $\phi\alpha_i\beta_j$. Furthermore, it makes the assumption
that the interaction effect is multiplicative in form. We have no particular reason to
believe it takes this form and so the alternative hypothesis may not be appropriate.
We execute the test:

```
> coef(g)
(Intercept)      laser50W      laser60W   tapemedium       tapefast
    23.9178        6.5733       12.2133       4.0367        -1.4800
> lasercoefs <- rep(c(0,6.5733,12.2133),3)
> tapecoefs <- rep(c(0,4.0367,-1.4800),each=3)
> h <- update(g, . ~ . + I(lasercoefs*tapecoefs))
> anova(h)
```

```
Analysis of Variance Table

Response: strength
                             Df Sum Sq Mean Sq F value Pr(>F)
laser                         2  224.2   112.1   36.82 0.0077
tape                          2   48.9    24.5    8.03 0.0624
I(lasercoefs * tapecoefs)     1    1.4     1.4    0.45 0.5503
Residuals                     3    9.1     3.0
```

The p-value of 0.55 indicates a nonsignificant interaction. So for these data, we can reasonably assume $(\alpha\beta)_{ij} = 0$. We can now test the main effects:

```
> anova(g)
Analysis of Variance Table

Response: strength
          Df Sum Sq Mean Sq F value Pr(>F)
laser      2  224.2   112.1   42.69  0.002
tape       2   48.9    24.5    9.32  0.031
Residuals  4   10.5     2.6
```

We see that both factors are significant. Examining the coefficients above, we see that the strength increases with the laser power but the strength is largest for medium tape speed but less for slow or fast tape speeds.

The treatment coding does not take advantage of the ordered nature of both factors. We can declare both to be *ordered factors* and refit:

```
> composite$laser <- as.ordered(composite$laser)
> composite$tape <- as.ordered(composite$tape)
> g <- lm(strength ~ laser + tape, composite)
> summary(g)

Coefficients:
              Estimate Std. Error t value Pr(>|t|)
(Intercept)     31.032      0.540   57.45 5.5e-07
laser.L          8.636      0.936    9.23 0.00077
laser.Q         -0.381      0.936   -0.41 0.70466
tape.L          -1.047      0.936   -1.12 0.32594
tape.Q          -3.900      0.936   -4.17 0.01404

Residual standard error: 1.62 on 4 degrees of freedom
Multiple R-Squared: 0.963,         Adjusted R-squared: 0.926
F-statistic:   26 on 4 and 4 DF,  p-value: 0.00401
```

Instead of a coding with respect to a reference level, we have linear and quadratic terms for each factor. The coding is:

```
> contr.poly(3)
               .L          .Q
[1,] -7.0711e-01    0.40825
[2,] -9.0733e-17   -0.81650
[3,]  7.0711e-01    0.40825
```

We see the linear term is proportional to $(-1, 0, 1)$ representing a linear trend across the levels while the quadratic term is proportional to $(1, -2, 1)$ representing a quadratic trend.

We see that the quadratic term for laser power is not significant while there is a quadratic effect for tape speed. One of the drawbacks of a model with factors is the difficulty of extrapolating to new conditions. The information gained from the ordered factors suggests a model with numerical predictors corresponding to the level values:

```
> composite$Ntape <- rep(c(6.42,13,27),each=3)
> composite$Nlaser <- rep(c(40,50,60),3)
> g1 <- lm(strength ~ Nlaser + poly(log(Ntape),2), composite)
> summary(g1)
```

```
Coefficients:
                       Estimate Std. Error t value Pr(>|t|)
(Intercept)              0.4989     3.0592    0.16  0.87684
Nlaser                   0.6107     0.0604   10.11  0.00016
poly(log(Ntape), 2)1    -1.8814     1.4791   -1.27  0.25933
poly(log(Ntape), 2)2    -6.7364     1.4791   -4.55  0.00609
```

```
Residual standard error: 1.48 on 5 degrees of freedom
Multiple R-Squared: 0.961,       Adjusted R-squared: 0.938
F-statistic: 41.5 on 3 and 5 DF,  p-value: 0.000587
```

We use the log of tape speed, as this results in roughly evenly spaced levels. This model fits about as well as the two-factor model but has the advantage that we make predictions for values of tape speed and laser power that were not used in the experiment. The earlier analysis with just factors helped us discover this model, which we may not otherwise have found.

15.3 Two-Way ANOVA with More than One Observation per Cell

Consider the case when the number of observations per cell is the same and greater than one, so that $n_{ij} = n > 1$ for all i, j. Such a design is orthogonal. Orthogonality can also occur if the row/column cell numbers are proportional.

With more than one observation per cell, we are now free to fit and test the model:

$$y_{ijk} = \mu + \alpha_i + \beta_j + (\alpha\beta)_{ij} + \varepsilon_{ijk}$$

The interaction effect may be tested by comparison to the model:

$$y_{ijk} = \mu + \alpha_i + \beta_j + \varepsilon_{ijk}$$

and computing the usual F-test. If the interaction effect is found to be significant, do not test the main effects even if they appear not to be significant. The estimation of the main effects and their significance is coding dependent when interactions are included in the model.

If the interaction effect is found to be insignificant, then test the main effects, but use RSS/df from the full model in the denominator of the F-tests — this has been shown to maintain the type I error better. So the F-statistic used is:

$$F = \frac{(RSS_{small} - RSS_{large})/(df_{small} - df_{large})}{\hat{\sigma}^2_{full}}$$

In an experiment to study factors affecting the production of the plastic polyvinyl chloride (PVC), three operators used eight different devices called resin railcars to produce PVC. For each of the 24 combinations, two samples were produced. The response is the particle size of the product. The experiment is described in Morris and Watson (1998).

We make some plots, as seen in Figure 15.3:

```
> data(pvc)
> attach(pvc)
> stripchart(psize ~ resin,xlab="Particle size",ylab="Resin railcar")
> stripchart(psize ~ operator,xlab="Particle size",ylab="Operator")
```

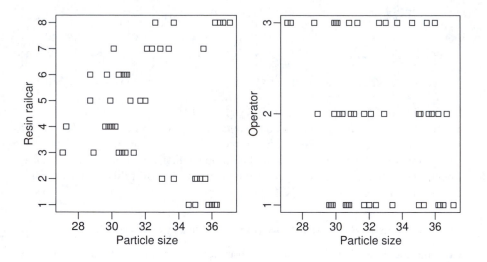

Figure 15.3 *PVC data plotted by resin railcars on the left and by operators on the right.*

Is there an interaction between operators and resin railcars? We first look at this graphically with interaction plots, as seen in Figure 15.4:

```
> interaction.plot(operator,resin,psize)
> interaction.plot(resin,operator,psize)
```

These are approximately parallel. The trouble with interaction plots is that we always expect there to be some random variation so it is sometimes difficult to distinguish true interaction from just noise. Fortunately, in this case, we have replication so we can directly test for an interaction effect.

Now fit the full model and see the significance of the factors:

```
> g <- lm(psize ~ operator*resin)
> anova(g)
```

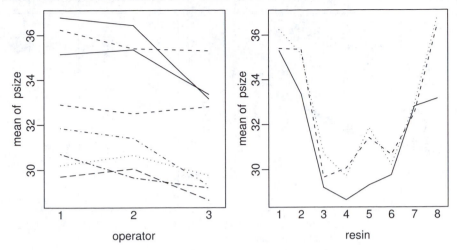

Figure 15.4 *Interaction plots for the PVC data.*

```
Analysis of Variance Table

Response: psize
              Df Sum Sq Mean Sq F value   Pr(>F)
operator       2   20.7    10.4    7.01    0.004
resin          7  283.9    40.6   27.44  5.7e-10
operator:resin 14   14.3     1.0    0.69    0.760
Residuals      24   35.5     1.5
```

We see that the interaction effect is not significant, but the main effects are. If you remove the interaction and then retest the main effects, you get a somewhat different result. As discussed earlier, the former test is preferred (although in this particular example, it does not affect the conclusion):

```
> anova(lm(psize ~ operator + resin, pvc))
Analysis of Variance Table

Response: psize
            Df Sum Sq Mean Sq F value   Pr(>F)
operator     2   20.7    10.4    7.9   0.0014
resin        7  283.9    40.6   30.9  8.1e-14
Residuals   38   49.8     1.3
```

We check the diagnostics, as seen in Figure 15.5:

```
> qqnorm(residuals(g))
> qqline(residuals(g))
> plot(fitted(g),residuals(g),xlab="Fitted",ylab="Residuals")
```

We see some evidence of outliers. The symmetry in the residuals vs. fitted plot is because, for each combination of the factors, the mean of the two replicates is the fitted value. The two residuals for that cell will be mirror images. If we exclude the

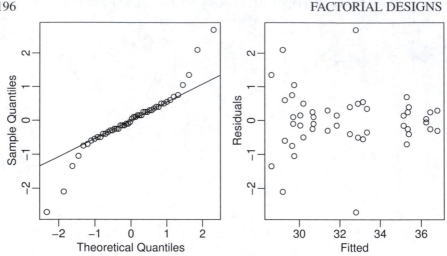

Figure 15.5 *Diagnostics plots for the full model for the PVC data.*

two largest outliers, then the interaction is still insignificant. An examination of the main effects model reveals that case 45 is the outlier, which we now exclude:

```
> g <- lm(psize ~ operator+resin,subset=-45)
> summary(g)
Coefficients:
              Estimate Std. Error t value Pr(>|t|)
(Intercept)     36.337      0.430   84.60  < 2e-16
operator2       -0.263      0.332   -0.79    0.435
operator3       -1.797      0.339   -5.30  5.5e-06
resin2          -1.033      0.543   -1.90    0.065
resin3          -5.800      0.543  -10.69  7.2e-13
resin4          -6.183      0.543  -11.40  1.2e-13
resin5          -4.800      0.543   -8.85  1.2e-10
resin6          -5.450      0.543  -10.04  4.1e-12
resin7          -3.692      0.570   -6.47  1.4e-07
resin8          -0.183      0.543   -0.34    0.737

Residual standard error: 0.94 on 37 degrees of freedom
Multiple R-Squared: 0.905,       Adjusted R-squared: 0.882
F-statistic: 39.2 on 9 and 37 DF,  p-value: 3.00e-16
```

The diagnostics for this model are satisfactory. We see that operator 1 and resin car 1 produce the largest particle size while operator 3 and resin car 4 produce the smallest.

We can construct pairwise confidence intervals for the treatment factor using the Tukey method:

```
> TukeyHSD(aov(psize ~ operator+resin,subset=-45))
  Tukey multiple comparisons of means
    95% family-wise confidence level
```

```
$operator
        diff      lwr       upr
2-1 -0.2625 -1.0737   0.54872
3-1 -1.7771 -2.6017  -0.95246
3-2 -1.5146 -2.3392  -0.68996

$resin
          diff        lwr        upr
2-1 -1.03333 -2.775204   0.70854
3-1 -5.80000 -7.541871  -4.05813
etc
```

We see that operators 1 and 2 are not significantly different but operator 3 is different from both. There are more differences in the resin cars.

The analysis above is appropriate for the investigation of specific operators and resin cars. These factors are being treated as *fixed effects*. If the operators and resin cars were randomly selected from larger populations of those available, they should be analyzed as *random effects*. This would require a somewhat different analysis not covered here. However, we can at least see from the analysis above that the variation between resin cars is greater than that between operators.

It is important that the observations taken in each cell are genuine replications. If this is not true, then the observations will be correlated and the analysis will need to be adjusted. It is a common scientific practice to repeat measurements and take the average to reduce measurement errors. These repeat measurements are not independent observations. Data where the replicates are correlated can be handled with repeated measures models. For example, in this experiment we would need to take some care to separate the two measurements for each operator and resin car. Some knowledge of the engineering might be necessary to achieve this.

15.4 Larger Factorial Experiments

Suppose we have factors $\alpha, \beta, \gamma, \ldots$ at levels $l_\alpha, l_\beta, l_\gamma, \ldots$. A *full* factorial experiment has at least one run for each combination of the levels. The number of combinations is $l_\alpha l_\beta l_\gamma \ldots$, which could easily be very large. The biggest model for a full factorial contains all possible interaction terms, which range from second-order, or two-way, as encountered earlier in this chapter, to high-order interactions involving several factors. For this reason, full factorials are rarely executed for more than three or four factors.

There are some advantages to factorial designs. If no interactions are significant, we get several one-way experiments for the price of one. Compare this with doing a sequence of one-way experiments. Also factorial experiments are efficient with experimental resources. It is often better to use replication for investigating another factor instead. For example, instead of doing a two-factor experiment with replication, it is often better to use that replication to investigate another factor.

The analysis of full factorial experiments is an extension of that used for the two-way ANOVA. Typically, there is no replication due to cost concerns so it is necessary to assume that some higher order interactions are zero in order to free up degrees of freedom for testing the lower order effects. Not many phenomena require a precise combination of several factors so this is not unreasonable.

Fractional factorials

Fractional factorials use only a fraction of the number of runs in a full factorial experiment. This is done to save the cost of the full experiment or to make only a few runs because the experimental material is limited. It is often possible to estimate the lower order effects with just a fraction. Consider an experiment with seven factors, each at two levels:

Effect	mean	main	2-way	3-way	4	5	6	7
Number of parameters	1	7	21	35	35	21	7	1

Table 15.1 *Number of parameters in a two-level, seven-factor experiment.*

If we are going to assume that higher order interactions are negligible then we do not really need $2^7 = 128$ runs to estimate the remaining parameters. We could perform only eight runs and still be able to estimate the seven main effects, though none of the interactions. In this particular example, it is hard to find a design to estimate all the two-way interactions uniquely, without a large number of runs. The construction of good designs is a complex task. For example, see Hamada and Wu (2000) for more on this. A Latin square (see Section 16.2) where all predictors are considered as factors is another example of a fractional factorial.

In fractional factorial experiments, we try to estimate many parameters with as few data points as possible. This means there are often not many degrees of freedom left. We require that σ^2 be small; otherwise there will be little chance of distinguishing significant effects. Fractional factorials are popular in engineering applications where the experiment and materials can be tightly controlled. Fractional factorials are popular in product design because they allow for the screening of a large number of factors. Factors identified in a screening experiment can then be more closely investigated. In the social sciences and medicine, the experimental materials, often human or animal, are much less homogeneous and less controllable, so σ^2 tends to be relatively larger. In such cases, fractional factorials are of no value.

Let's look at an example. Speedometer cables can be noisy because of shrinkage in the plastic casing material. An experiment was conducted to find out what caused shrinkage by screening a large number of factors. The engineers started with 15 different factors: liner outside diameter, liner die, liner material, liner line speed, wire braid type, braiding tension, wire diameter, liner tension, liner temperature, coating material, coating die type, melt temperature, screen pack, cooling method and line speed, labeled a through o. Response is percentage of shrinkage per specimen. There were two levels of each factor. The "+" indicates the high level of a factor and the "−" indicates the low level.

A full factorial would take 2^{15} runs, which is highly impractical, thus a design with only 16 runs was used where the particular runs have been chosen specially so as to estimate the mean and the 15 main effects. We assume that there is no interaction effect of any kind. The data come from Box, Bisgaard, and Fung (1988).

Read in and check the data:

```
> data(speedo)
> speedo
   h d l b j f n a i e m c k g o      y
1  - - + - + + - - + + - + - - + 0.4850
2  + - - - - + + - - + + + + - - 0.5750
......
16 + + + + + + + + + + + + + + + 0.5825
```

Fit and examine a main effects only model:

```
> g <- lm(y ~ .,speedo)
> summary(g)
Residuals:
ALL 16 residuals are 0: no residual degrees of freedom!

Coefficients:
              Estimate Std. Error t value Pr(>|t|)
(Intercept)  0.5825000         NA      NA       NA
h-          -0.0621875         NA      NA       NA
d-          -0.0609375         NA      NA       NA
l-          -0.0271875         NA      NA       NA
b-           0.0559375         NA      NA       NA
j-           0.0009375         NA      NA       NA
f-          -0.0740625         NA      NA       NA
n-          -0.0065625         NA      NA       NA
a-          -0.0678125         NA      NA       NA
i-          -0.0428125         NA      NA       NA
e-          -0.2453125         NA      NA       NA
m-          -0.0278125         NA      NA       NA
c-          -0.0896875         NA      NA       NA
k-          -0.0684375         NA      NA       NA
g-           0.1403125         NA      NA       NA
o-          -0.0059375         NA      NA       NA

Residual standard error: NaN on 0 degrees of freedom
Multiple R-Squared:     1,       Adjusted R-squared:    NaN
F-statistic:   NaN on 15 and 0 DF,   p-value: NA
```

There are no degrees of freedom, because there are as many parameters as cases. We cannot do any of the usual tests. It is important to understand the coding here, so look at the X-matrix:

```
> model.matrix(g)
   (Intercept) h d l b j f n a i e m c k g o
1            1 1 1 0 1 0 0 1 1 0 0 1 0 1 1 0
...etc...
```

We see that "+" is coded as zero and "−" is coded as one. This unnatural ordering is because of the order of "+" and "−" in the ASCII alphabet.

We do not have any degrees of freedom so we can not make the usual F-tests. We
need a different method to determine significance. Suppose there were no significant
effects and the errors were normally distributed. The estimated effects would then
just be linear combinations of the errors and hence normal. We now make a normal quantile plot of the main effects with the idea that outliers represent significant
effects:

```
> qqnorm(coef(g)[-1],pch=names(coef(g)[-1]))
```

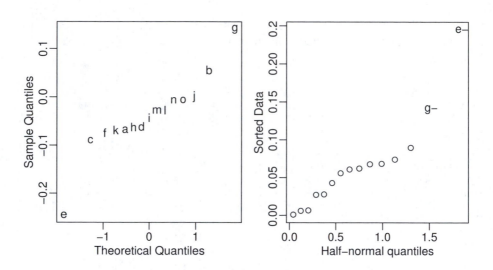

Figure 15.6 *Q–Q plots of effects for speedometer cable analysis.*

See Figure 15.6. Notice that "e" and possibly "g" are extreme. Since the "e" effect
is negative, the "+" level of "e" increases the response. Since shrinkage is a bad thing,
increasing the response is not good so we would prefer whatever "wire braid" type
corresponds to the "−" level of "e". The same reasoning for "g" leads us to expect
that a larger (assuming that is "+") would decrease shrinkage.

A half-normal plot is better for detecting extreme points:

```
> halfnorm(coef(g)[-1],labs=names(coef(g)[-1]))
```

We might now conduct another experiment focusing on the effect of "e" and "g."

Exercises

1. Analyze `warpbreaks` data as a two-way ANOVA. Which factors are significant? Now check for a good transformation on the response and see whether the
   model may be simplified. Now form a six-level factor from all combinations of
   the `wool` and `tension` factors. Which combinations are significantly different?

2. The `barley` data may be found in the `lattice` package. Perform a three-way
   ANOVA with `yield` as the response. Check the diagnostics — you will find that

two points stick out, which correspond to the same variety and site, but for different years. There is reason to suspect that the response values for these cases have been transposed. Investigate the effect of transposing these cases on the analysis.

3. Determine the important factors in the `sono` dataset where the `Intensity` is the response and the other variables are predictors.

4. Using the `rats` data, model the survival time in terms of the poison and treatment. Use the Box–Cox method to determine an appropriate transformation on the response. Which treatments and which poisons are significantly different?

5. The `peanut` data come from a fractional factorial experiment to investigate factors that affect an industrial process using carbon dioxide to extract oil from peanuts. Determine which factors are important remembering to consider two-way interactions.

CHAPTER 16

Block Designs

In a completely randomized design (CRD), the treatments are assigned to the experimental units at random. This is appropriate when the units are homogeneous, as has been assumed in the designs leading to the one- and two-way analysis of variances (ANOVAs). Sometimes, we may suspect that the units are heterogeneous, but we can not describe the form the difference takes — for example, we may know that a group of patients are not identical, but we may have no further information about them. In this case, it is still appropriate to use a CRD. Of course, the randomization will tend to spread the heterogeneity around to reduce bias, but the real justification lies in the randomization test discussed in Section 3.3. Under the null hypothesis, there is no link between a factor and the response. In other words, the responses have been assigned to the units in a way that is unlinked to the factor. This corresponds to the randomization used in assigning the levels of the factor to the units. This is why the randomization is crucial because it allows us to make this argument. Now if the difference in the response between levels of the factor seems too unlikely to have occurred by chance, we can reject the null hypothesis. The normal-based inference is approximately equivalent to the permutation-based test. Since the normal-based inference is much quicker, we might prefer to use that.

When the experimental units are heterogeneous in a known way and can be arranged into *blocks* where the within block variation is ideally small, but the between block variation is large, a *block design* can be more efficient than a CRD. We prefer to have a block size equal to the number of treatments. If this cannot be done, an *incomplete* block design must be used.

Sometimes the blocks are determined by the experimenter. For example, suppose we want to compare four treatments and have 20 patients available. We might divide the patients into five blocks of four patients each where the patients in each block have some relevant similarity. We might decide this subjectively in the absence of specific information. In other cases, the blocks are predetermined by the nature of the experiment. For example, suppose we want to test three crop varieties on five fields. Restrictions on planting, harvesting and irrigation equipment might allow us only to divide the fields into three strips.

In a randomized block design, the treatment levels are assigned randomly within a block. This means the randomization is restricted relative to the full randomization used in the CRD. This has consequences for the inference. There are fewer possible permutations for the random assignment of the treatments, therefore, the computation of the significance of a statistic based on the permutation test would need to be modified. Similarly, a block effect must be included in the model used for inference about the treatments, even if the block effect is not significant.

16.1 Randomized Block Design

We have one treatment factor, τ at t levels and one blocking factor, β at r levels. The model is:

$$y_{ij} = \mu + \tau_i + \beta_j + \varepsilon_{ij}$$

where τ_i is the treatment effect and ρ_j is the blocking effect. There is one observation on each treatment in each block. This is called a randomized complete block design (RCBD). The analysis is then very similar to the two-way ANOVA with one observation per cell. We can check for an interaction and check for a treatment effect. We can also check the block effect, but this is only useful for future reference. Blocking is a feature of the experimental units and restricts the randomized assignment of the treatments. This means that we cannot regain the degrees of freedom devoted to blocking even if the blocking effect is not significant. The randomization test-based argument means that we must judge the magnitude of the treatment effect within the context of the restricted randomization that has been used.

We illustrate this with an experiment to compare eight varieties of oats. The growing area was heterogeneous and so was grouped into five blocks. Each variety was sown once within each block and the yield in grams per 16-ft row was recorded. The data come from Anderson and Bancroft (1952).

We start with a look at the data:

```
> data(oatvar)
> attach(oatvar)
> xtabs(yield ~ variety + block)
        block
variety I    II   III IV   V
      1 296  357  340 331  348
      2 402  390  431 340  320
      3 437  334  426 320  296
      4 303  319  310 260  242
      5 469  405  442 487  394
      6 345  342  358 300  308
      7 324  339  357 352  220
      8 488  374  401 338  320
> stripchart(yield ~ variety,xlab="yield",ylab="variety")
> stripchart(yield ~ block,xlab="yield",ylab="block")
```

See Figure 16.1. There is no indication of outliers, skewness or nonconstant variance. Now check for interactions, as seen in Figure 16.2:

```
> interaction.plot(variety,block,yield)
> interaction.plot(block,variety,yield)
```

There is no clear evidence of interaction:

```
> g <- lm(yield ~ block+variety)
> anova(g)
Analysis of Variance Table

Response: yield
          Df Sum Sq Mean Sq F value  Pr(>F)
```

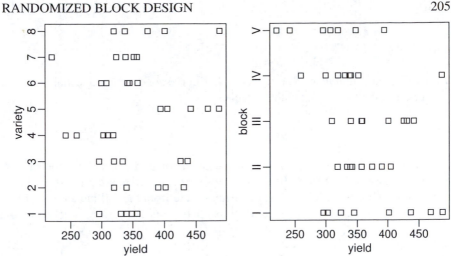

Figure 16.1 *Strip plots of oat variety data.*

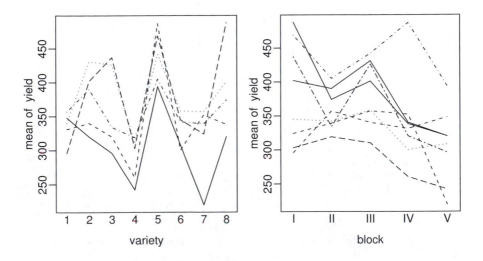

Figure 16.2 *Interaction plots for oat variety data.*

```
block        4   33396      8349      6.24    0.001
variety      7   77524     11075      8.28 1.8e-05
Residuals 28     37433      1337
```

Both effects are significant. The ANOVA table corresponds to a sequential testing of models, here corresponding to the sequence:

```
y ~ 1
y ~ block
y ~ block+variety
```

So here the p-value 0.001 corresponds to a comparison of the first two models in this list, while the p-value of 1.8e-05 corresponds to the test comparing the second two. The denominator in both F-tests is the mean square from the full model, here 1337. This means that a test of the block effect that leaves out the variety effect is not the same:

```
> anova(lm(yield ~ block))
Analysis of Variance Table

Response: yield
          Df Sum Sq Mean Sq F value Pr(>F)
block      4  33396    8349    2.54  0.057
Residuals 35 114957    3284
```

There is a difference in significance in this case. This latter test is incorrect for testing the blocking effect.

Notice that if we change the order of the terms in the ANOVA, it makes no difference because of the orthogonal design:

```
> anova(lm(yield ~ variety+block))
Analysis of Variance Table

Response: yield
          Df Sum Sq Mean Sq F value  Pr(>F)
variety    7  77524   11075    8.28 1.8e-05
block      4  33396    8349    6.24   0.001
Residuals 28  37433    1337
```

By way of comparison, see what happens if we omit the first observation in the dataset — this might happen in practice if this run is lost:

```
> anova(lm(yield ~ block+variety,subset=-1))
Analysis of Variance Table

Response: yield
          Df Sum Sq Mean Sq F value  Pr(>F)
block      4  38581    9645    8.41 0.00015
variety    7  75339   10763    9.38 7.3e-06
Residuals 27  30968    1147
> anova(lm(yield ~ variety+block,subset=-1))
Analysis of Variance Table

Response: yield
```

```
           Df Sum Sq Mean Sq F value  Pr(>F)
variety     7  75902   10843    9.45 6.8e-06
block       4  38018    9504    8.29 0.00017
Residuals 27  30968    1147
```

As there is one missing observation, the design is no longer orthogonal and the order does matter, although it would not change the general conclusions. If we want to test for a treatment effect, we would prefer the first of these two tables since in that version the blocking factor is already included when we test the treatment factor. Since the blocking factor is an unalterable feature of the chosen design, this is as it should be. A convenient way to test all terms relative to the full model is:

```
> drop1(lm(yield ~ variety+block,subset=-1),test="F")
Single term deletions

Model:
yield ~ variety + block
        Df Sum of Sq    RSS    AIC F value    Pr(F)
<none>                 30968    284
variety  7     75339 106307    319    9.38 7.3e-06
block    4     38018  68986    308    8.29 0.00017
```

Check the diagnostics, as seen in Figure 16.3:

```
> plot(fitted(g),residuals(g),xlab="Fitted",ylab="Residuals")
> abline(h=0)
> qqnorm(residuals(g))
> qqline(residuals(g))
```

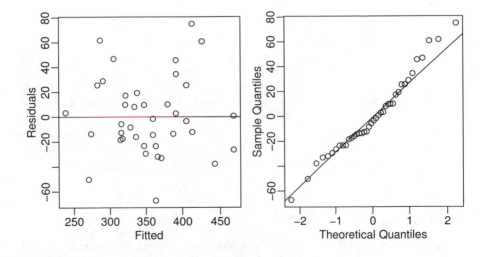

Figure 16.3 *Diagnostic plots for the oat variety data.*

An examination of which varieties give the highest yields and which are significantly better than others can now follow.

We did assume that the interactions were not significant. We looked at the interaction plots, but we can also execute the Tukey nonadditivity test:

```
> varcoefs <-  c(0,coef(g)[6:12])
> blockcoefs <-c(0,coef(g)[2:5])
> ab <- rep(varcoefs,each=5)*rep(blockcoefs,8)
> h <- update(g,.~.+ab)
> anova(h)
Analysis of Variance Table

Response: yield
          Df Sum Sq Mean Sq F value  Pr(>F)
block      4  33396    8349    6.06  0.0013
variety    7  77524   11075    8.03 2.8e-05
ab         1    213     213    0.15  0.6974
Residuals 27  37220    1379
```

Because the p-value of the treatment times block effect is 0.6974, we accept the null hypothesis of no interaction. Of course, the interaction may be of a nonmultiplicative form, but there is little we can do about that.

Relative advantage of RCBD over CRD

We can measure precision by considering var $\hat{\tau}$ or equivalently $\hat{\sigma}^2$. We should compare the $\hat{\sigma}^2$ for designs with the same sample size. We define *relative efficiency* as $\hat{\sigma}^2_{CRD}/\hat{\sigma}^2_{RCBD}$ where the quantities can be computed by fitting models with and without the blocking effect. For the example above:

```
> gcrd <- lm(yield ~ variety)
> summary(gcrd)$sig
[1] 47.047
> summary(g)$sig
[1] 33.867
> (47.047/33.867)^2
[1] 1.9298
```

So a CRD would require 93% more observations to obtain the same level of precision as an RCBD.

The efficiency is not guaranteed to be greater than one. Only use blocking where there is some heterogeneity in the experimental units. The decision to block is a matter of judgment prior to the experiment. There is no guarantee that it will increase precision.

16.2 Latin Squares

Latin squares are useful when there are two blocking variables. For example, in a field used for agricultural experiments, the level of moisture may vary across the field in one direction and the fertility, in another. In an industrial experiment, suppose we wish to compare four production methods (the treatments) — A, B, C and D. We have available four machines 1, 2, 3 and 4, and four operators, I, II, III and IV. A Latin square design is shown in Table 16.1.

	1	2	3	4
I	A	B	C	D
II	B	D	A	C
III	C	A	D	B
IV	D	C	B	A

Table 16.1 *Latin square showing the treatment (A to D) used for different combinations of two factors.*

Each treatment is assigned to each block once and only once. We should choose randomly from all the possible Latin square layouts.

Let τ be the treatment factor and β and γ be the two blocking factors; then the model is:

$$y_{ijk} = \mu + \tau_i + \beta_j + \gamma_k + \varepsilon_{ijk} \qquad i, j, k = 1, \ldots, t$$

All combinations of i, j and k do not appear. To test for a treatment effect simply fit a model without the treatment effect and compare using the F-test. The Tukey pairwise CIs are:

$$\hat{\tau}_l - \hat{\tau}_m \pm \frac{q_{t,(t-1)(t-2)}}{\sqrt{2}} \hat{\sigma} \sqrt{2/t}$$

The Latin square can be even more efficient than the RCBD provided that the blocking effects are sizable. There are some variations on the Latin square. The Latin square can be replicated if more runs are available. We need to have both block sizes to be equal to the number of treatments. This may be difficult to achieve. Latin rectangle designs are possible by adjoining Latin squares. When there are three blocking variables, a Graeco–Latin square may be used but these rarely arise in practice.

The Latin square can also be used for comparing three treatment factors. Only t^2 runs are required compared to the t^3 required if all combinations were run. (The downside is that you can not estimate the interactions if they exist.) This is an example of a *fractional factorial*.

In an experiment reported by Davies (1954), four materials, A, B, C and D, were fed into a wear-testing machine. The response is the loss of weight in 0.1 mm over the testing period. The machine could process four samples at a time and past experience indicated that there were some differences due to the position of these four samples. Also some differences were suspected from run to run. Four runs were made:

```
> data(abrasion)
> abrasion
   run position material wear
1    1        1         C  235
2    1        2         D  236
..etc..
16   4        4         D  225
```

We can check the Latin square structure:

```
> matrix(abrasion$material,4,4)
     [,1] [,2] [,3] [,4]
```

```
[1,] "C"  "A"  "D"  "B"
[2,] "D"  "B"  "C"  "A"
[3,] "B"  "D"  "A"  "C"
[4,] "A"  "C"  "B"  "D"
```

Plot the data:

```
> with(abrasion,stripchart(wear ~ material,
  xlab="Material",vert=T))
> with(abrasion,stripchart(wear ~ run,xlab="Run",vert=T))
> with(abrasion,stripchart(wear ~ position,
  xlab="Position",vert=T))
```

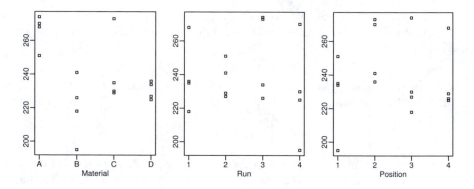

Figure 16.4 *Amount of wear depending on material, run and position.*

Examine the plots in Figure 16.4. There appear to be differences in all variables. No outliers, skewness or unequal variance is apparent.

Now fit the Latin square model and test each variable relative to the full model:

```
> g <- lm(wear ~ material+run+position, abrasion)
> drop1(g,test="F")
Single term deletions

Model:
wear ~ material + run + position
          Df Sum of Sq  RSS  AIC F value    Pr(F)
<none>                  368   70
material   3      4621 4989  106   25.15  0.00085
run        3       986 1354   85    5.37  0.03901
position   3      1468 1836   90    7.99  0.01617
```

We see that all variables are statistically significant. There are clear differences between the materials. We checked the diagnostics on the model, which showed nothing remarkable. We examine the coefficients:

```
> summary(g)

Coefficients:
```

```
              Estimate Std.  Error t value Pr(>|t|)
(Intercept)     254.75        6.19   41.17 1.4e-08
materialB       -45.75        5.53   -8.27 0.00017
materialC       -24.00        5.53   -4.34 0.00489
materialD       -35.25        5.53   -6.37 0.00070
run2             -2.25        5.53   -0.41 0.69842
run3             12.50        5.53    2.26 0.06466
run4             -9.25        5.53   -1.67 0.14566
position2        26.25        5.53    4.74 0.00318
position3         8.50        5.53    1.54 0.17545
position4         8.25        5.53    1.49 0.18661
```

```
Residual standard error: 7.83 on 6 degrees of freedom
Multiple R-Squared: 0.951,        Adjusted R-squared: 0.877
F-statistic: 12.8 on 9 and 6 DF,  p-value: 0.00283
```

We see that material B looks best (in terms of least wear) followed by material D. Is the difference significant though? Which materials in general are significantly better than others? We need the Tukey pairwise intervals to help determine this. The width of the band is calculated in the usual manner:

```
> qtukey(0.95,4,6)*5.53/sqrt(2)
[1] 19.143
```

The width of the interval is 19.1. We can make a table of the material differences:

```
> scoefs <- c(0,coef(g)[2:4])
> outer(scoefs,scoefs,"-")
                  materialB materialC materialD
            0.00     45.75     24.00     35.25
materialB -45.75      0.00    -21.75    -10.50
materialC -24.00     21.75      0.00     11.25
materialD -35.25     10.50    -11.25      0.00
```

We see that the (B, D) and (D, C) differences are not significant at the 5% level, but that all the other differences are significant.

If maximizing resistance to wear is our aim, we would pick material B but if material D offered a better price, we might have some cause to consider switching to D. The decision would need to be made with cost–quality trade-offs in mind.

Now we compute how efficient the Latin square is compared to other designs. We compare to the completely randomized design:

```
> gr <- lm(wear ~ material,abrasion)
> (summary(gr)$sig/summary(g)$sig)^2
[1] 3.8401
```

We see that the Latin square is 3.84 times more efficient than the CRD. This is a substantial gain in efficiency. The Latin square may also be combined to designs where we block on only one of the variables. The efficiency relative to these designs is less impressive, but still worthwhile.

16.3 Balanced Incomplete Block Design

In a complete block design, the block size is equal to the number of treatments. When the block size is less than the number of treatments, an incomplete block design must be used. For example, in the oat example, suppose six oat varieties were to be compared, but each field had space for only four plots.

In an incomplete block design, the treatments and blocks are *not* orthogonal. Some treatment contrasts will not be identifiable from certain block contrasts. This is an example of *confounding*. This means that those treatment contrasts effectively cannot be examined. In a balanced incomplete block (BIB) design, all the pairwise differences are identifiable and have the same standard error. Pairwise differences are more likely to be interesting than other contrasts, so the design is constructed to facilitate this.

Suppose, we have four treatments ($t = 4$) A,B,C,D and the block size, $k = 3$ and there are $b = 4$ blocks. Therefore, each treatment appears $r = 3$ times in the design. One possible BIB design is:

Block 1	A	B	C
Block 2	A	B	D
Block 3	A	C	D
Block 4	B	C	D

Table 16.2 *BIB design for four treatments with four blocks of size three.*

Each pair of treatments appears in the same block $\lambda = 2$ times — this feature means simpler pairwise comparison is possible. For a BIB design, we require:

$$
\begin{aligned}
b &\geq\ t > k \\
rt &=\ bk = n \\
\lambda(t-1) &=\ r(k-1)
\end{aligned}
$$

This last relation holds because the number of pairs in a block is $k(k-1)/2$ so the total number of pairs must be $bk(k-1)/2$. On the other hand, the number of treatment pairs is $t(t-1)/2$. The ratio of these two quantities must be λ.

Since λ has to be an integer, a BIB design is not always possible even when the first two conditions are satisfied. For example, consider $r = 4, t = 3, b = 6, k = 2$ and then $\lambda = 2$ which is OK, but if $r = 4, t = 4, b = 8, k = 2$, then $\lambda = 4/3$ so no BIB is possible. (Something called a partially balanced incomplete block design can then be used.) BIBs are also useful for competitions where not all contestants can fit in the same race.

The model we fit is the same as for the RCBD:

$$
y_{ij} = \mu + \tau_i + \beta_j + \varepsilon_{ij}
$$

In our example, a nutritionist studied the effects of six diets, "a" through "f," on weight gain of domestic rabbits. From past experience with sizes of litters, it was felt that only three uniform rabbits could be selected from each available litter. There were ten litters available forming blocks of size three. The data come from Lentner and Bishop (1986). Examine the data:

```
> data(rabbit)
> xtabs(gain ~ treat+block, rabbit)
     block
treat b1    b10  b2   b3   b4   b5   b6   b7   b8   b9
    a  0.0 37.3 40.1  0.0 44.9  0.0  0.0 45.2 44.0  0.0
    b 32.6  0.0 38.1  0.0  0.0  0.0 37.3 40.6  0.0 30.6
    c 35.2  0.0 40.9 34.6 43.9 40.9  0.0  0.0  0.0  0.0
    d  0.0 42.3  0.0 37.5  0.0 37.3  0.0 37.9  0.0 27.5
    e  0.0  0.0  0.0  0.0 40.8 32.0 40.5  0.0 38.5 20.6
    f 42.2 41.7  0.0 34.3  0.0  0.0 42.8  0.0 51.9  0.0
```

The zero values correspond to no observation. The BIB structure is apparent — each pair of diets appear in the same block exactly twice. Now plot the data, as seen in Figure 16.5:

```
> attach(rabbit)
> stripchart(gain ~ block,xlab="weight gain",ylab="block")
> stripchart(gain ~ treat,xlab="weight gain",ylab="treat")
```

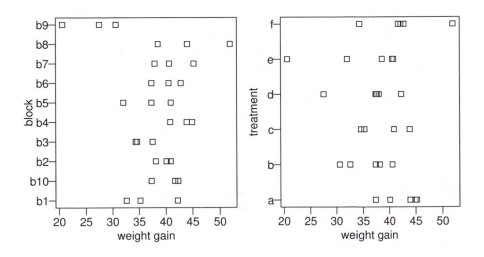

Figure 16.5 *Strip plots of rabbit diet data.*

We fit the model taking care to put the blocking factor first:

```
> g <- lm(gain ~ block+treat,rabbit)
> anova(g)
Analysis of Variance Table

Response: gain
          Df Sum Sq Mean Sq F value  Pr(>F)
block      9    730      81    8.07 0.00025
```

```
treat        5      159      32     3.16 0.03817
Residuals 15      151      10
```

Changing the order of treatment and block:

```
> anova(lm(gain ~ treat+block,rabbit))
Analysis of Variance Table

Response: gain
          Df Sum Sq Mean Sq F value  Pr(>F)
treat      5    293      59    5.84 0.00345
block      9    596      66    6.59 0.00076
Residuals 15    151      10
```

does make a difference because the design is not orthogonal because of the incompleteness. We prefer the first table, because we want to test for a treatment effect after the blocking effect has been considered.

Now check the diagnostics:

```
> plot(fitted(g),residuals(g),xlab="Fitted",ylab="Residuals")
> abline(h=0)
> qqnorm(residuals(g),main="")
> qqline(residuals(g))
```

The plots are not shown, as they show nothing of interest.

Now we check which treatments differ. The Tukey pairwise CIs need to be directly constructed because this is not a complete layout. We extract the information about the treatment effects and the standard error:

```
> coef(summary(g))
          Estimate Std. Error   t value Pr(>|t|)
treatb  -1.741667     2.2418 -0.776898 0.449298
treatc   0.400000     2.2418  0.178426 0.860776
treatd   0.066667     2.2418  0.029738 0.976668
treate  -5.225000     2.2418 -2.330695 0.034134
treatf   3.300000     2.2418  1.472018 0.161686
```

We see that the standard error for the pairwise comparison is 2.24. Notice that all the treatment standard errors are equal because of the BIB. Now compute the Tukey critical value:

```
> qtukey(0.95,6,15)
[1] 4.5947
```

So the intervals have width:

```
> 4.59*2.24/sqrt(2)
[1] 7.2702
```

We check which pairs are significantly different:

```
> tcoefs <- c(0,coef(g)[11:15])
> abs(outer(tcoefs,tcoefs,"-")) > 7.27
                treatb treatc treatd treate treatf
        FALSE    FALSE  FALSE  FALSE  FALSE  FALSE
treatb  FALSE    FALSE  FALSE  FALSE  FALSE  FALSE
treatc  FALSE    FALSE  FALSE  FALSE  FALSE  FALSE
```

```
treatd FALSE    FALSE    FALSE    FALSE    FALSE    FALSE
treate FALSE    FALSE    FALSE    FALSE    FALSE     TRUE
treatf FALSE    FALSE    FALSE    FALSE     TRUE    FALSE
```

Only the $e - f$ difference is significant.

Now let's see how much better this blocked design is than the CRD. We compute the relative efficiency:

```
> gr <- lm(gain ~ treat,rabbit)
> (summary(gr)$sig/summary(g)$sig)^2
[1] 3.0945
```

Blocking was well worthwhile here.

Exercises

1. The alfalfa data arise from a Latin square design where the treatment factor is inoculum and the blocking factors are shade and irrigation. Test the significance of the effects and determine which levels of the treatment factor are significantly different.

2. The eggprod comes from a randomized block experiment to determine factors affecting egg production. Is there a difference in the treatments? What is the nature of the difference? What efficiency was gained by the blocked design?

3. The morley data can be viewed as a randomized block experiment with Run as the treatment factor and Expt as the blocking factor. Is there a difference between runs and what efficiency is gained by blocking?

4. The OrchardSprays data arise from a Latin square design. Which factors show a difference? What pairs of treatments are different? Can the treatment factor be replaced with a linear term? What efficiency is gained by this design?

5. The resceram data arise from an experiment to test the effects of resistor shape on current noise. The resistors are mounted on plates and only three resistors will fit on a plate although there are four different shapes to be tested. Identify the design and analyze. Which shapes are different?

R Installation, Functions and Data

R may be obtained from the R project Web site at `www.r-project.org`.

This book uses some functions and data that are not part of base R. You may wish to download these extras from the R Web site. The additional packages used are:

`MASS, nlme, splines, leaps, quantreg, lmtest, pls, ellipse, faraway`

`MASS, splines` and `nlme` are part of the "recommended" R installation; you will have these already unless you choose a nonstandard installation. Use the command:

```
> library()
```

within R to see what packages you have. Under Windows, to install the additional packages, choose the "Install packages from CRAN" menu option. You must have a network connection for this to work — if you are working offline, you may use the "Install packages from local zip file" menu option provided you have already obtained the necessary packages. Under other operating systems, such as Macintosh or Linux, the installation procedure differs. Consult the R Web site for details.

I have packaged the data and functions that I have used in this book as an R package called `faraway` that you may obtain from CRAN or the book Web site at `www.maths.bath.ac.uk/~jjf23/LMR`. The functions defined are:

halfnorm	Half normal plot
qqnorml	Case-labeled Q-Q plot
vif	Variance Inflation factors
prplot	Partial residual plot

In addition the following datasets are used as examples in the text:

abrasion	Wear test experiment with Latin square design
chredlin	Chicago insurance redlining
chmiss	Chicago data with some missing values
coagulation	Blood coagulation times by diet
composite	Strength of a thermoplastic composite depending on two factors
corrosion	Corrosion loss in Cu-Ni alloys
eco	Ecological regression example
fpe	1981 French presidential election
fruitfly	Longevity of fruitflies depending on sexual activity
gala	Species diversity on the Galapagos Islands
meatspec	Meat spectrometry to determine fat content
oatvar	Yields of oat varieties grown in blocks
odor	Odor of chemical by production settings

pima	Diabetes survey on Pima Indians
pvc	Production of PVC by operator and resin railcar
rabbit	Rabbit weight gain by diet and litter
savings	Savings rates in 50 countries
seatpos	Car seat position depending driver size
sexab	Post traumatic stress disorder in abused women
speedo	Speedometer cable shrinkage
star	Star light intensities and temperatures
stat500	Scores for students in Stat500 class

Some additional datasets are provided for the exercises.

Where add-on packages are needed in the text, you will find the appropriate `library()` command. However, I have assumed that the `faraway` library is always loaded. You can add a line reading `library(faraway)` to your Rprofile file if you expect to use this package in every session. Otherwise, you will need to remember to type it each time.

I set the following options to achieve the output seen in this book:

```
> options(digits=5,show.signif.stars=FALSE)
```

The `digits=5` reduces the number of digits shown when printing numbers from the default of seven. Note that this does not reduce the precision with which these numbers are internally stored. One might take this further — anything more than two or three significant digits in a displayed table is usually unnecessary and more importantly, distracting. I have also edited the output in the text to remove extraneous output or to improve the formatting.

The code and output shown in this book were generated under R version 1.9.0. R is regularly updated and improved so more recent versions may show some differences in the output.

Quick Introduction to R

This is just a brief introduction to R. See the preface for recommendations about how to learn more about R.

B.1 Reading the Data In

The first step is to read the data in. You can use the `read.table()` or `scan()` functions to read data in from outside R. You can also use the `data()` function to access data already available within R:

```
> data(stackloss)
> stackloss
   Air.Flow Water.Temp Acid.Conc. stack.loss
1       80         27         89         42
2       80         27         88         37
... stuff deleted ...
21      70         20         91         15
```

Type:

```
> help(stackloss)
```

to see more information about the data. We can check the dimension of the data:

```
> dim(stackloss)
[1] 21   4
```

There are 21 rows and four columns.

B.2 Numerical Summaries

One easy way to get the basic numerical summaries is:

```
> summary(stackloss)
    Air.Flow        Water.Temp       Acid.Conc.       stack.loss
 Min.   :50.0    Min.   :17.0     Min.   :72.0     Min.   : 7.0
 1st Qu.:56.0    1st Qu.:18.0     1st Qu.:82.0     1st Qu.:11.0
 Median :58.0    Median :20.0     Median :87.0     Median :15.0
 Mean   :60.4    Mean   :21.1     Mean   :86.3     Mean   :17.5
 3rd Qu.:62.0    3rd Qu.:24.0     3rd Qu.:89.0     3rd Qu.:19.0
 Max.   :80.0    Max.   :27.0     Max.   :93.0     Max.   :42.0
```

We can also compute these numbers separately:

```
> stackloss$Air.Flow
 [1] 80 80 75 62 62 62 62 62 58 58 58 58 58 58 50 50 50 50 50 56
[21] 70
```

```
> mean(stackloss$Ai)
[1] 60.429
> median(stackloss$Ai)
[1] 58
> range(stackloss$Ai)
[1] 50 80
> quantile(stackloss$Ai)
  0%  25%  50%  75% 100%
  50   56   58   62   80
```

We can get the variance and standard deviation:

```
> var(stackloss$Ai)
[1] 84.057
> sd(stackloss$Ai)
[1] 9.1683
```

We might also want the correlations:

```
> cor(stackloss)
            Air.Flow Water.Temp Acid.Conc.  stack.loss
Air.Flow     1.00000    0.78185    0.50014     0.91966
Water.Temp   0.78185    1.00000    0.39094     0.87550
Acid.Conc.   0.50014    0.39094    1.00000     0.39983
stack.loss   0.91966    0.87550    0.39983     1.00000
```

B.3 Graphical Summaries

We can make histograms and boxplots and specify the labels:

```
> hist(stackloss$Ai)
> hist(stackloss$Ai,main="Histogram of Air Flow",
  xlab="Flow of cooling air")
> boxplot(stackloss$Ai)
```

Scatterplots are also easily constructed:

```
> plot(stackloss$Ai,stackloss$W)
> plot(Water.Temp ~ Air.Flow,stackloss,xlab="Air Flow",
  ylab="Water Temperature")
```

We can make a scatterplot matrix:

```
> plot(stackloss)
```

We can put several plots in one display:

```
> par(mfrow=c(2,2))
> boxplot(stackloss$Ai)
> boxplot(stackloss$Wa)
> boxplot(stackloss$Ac)
> boxplot(stackloss$s)
> par(mfrow=c(1,1))
```

where the final command causes a return to one plot per display.

B.4 Selecting Subsets of the Data

The second row:

```
> stackloss[2,]
  Air.Flow Water.Temp Acid.Conc. stack.loss
2       80         27         88          37
```

The third column:

```
> stackloss[,3]
 [1] 89 88 90 87 87 87 93 93 87 80 89 88 82 93 89 86 72 79 80 82
[21] 91
```

The (2, 3) element:

```
> stackloss[2,3]
[1] 88
```

c() is a function for making vectors — for example:

```
> c(1,2,4)
[1] 1 2 4
```

Select the first, second and fourth rows:

```
> stackloss[c(1,2,4),]
  Air.Flow Water.Temp Acid.Conc. stack.loss
1       80         27         89          42
2       80         27         88          37
4       62         24         87          28
```

The : operator is good for making sequences — for example:

```
> 3:11
 [1]  3  4  5  6  7  8  9 10 11
```

We can select the third through sixth rows:

```
> stackloss[3:6,]
  Air.Flow Water.Temp Acid.Conc. stack.loss
3       75         25         90          37
4       62         24         87          28
5       62         22         87          18
6       62         23         87          18
```

We can use "−" to indicate "everything but" — that is all the data except the first two columns:

```
> stackloss[,-c(1,2)]
  Acid.Conc. stack.loss
1         89         42
2         88         37
...
21        91         15
```

We may also want to select the subsets on the basis of some criterion — for example, those cases which have an air flow greater than 72:

```
> stackloss[stackloss$Ai > 72,]
  Air.Flow Water.Temp Acid.Conc. stack.loss
1       80         27         89          42
2       80         27         88          37
3       75         25         90          37
```

B.5 Learning More about R

While running R you can get help about a particular command, for example, if you want help about the `boxplot()` command, just type `help(boxplot)`. If you do not know what the name of the command is that you want to use, then type:

```
help.start()
```

and then browse. You will be able to pick up the language from the examples in the text and from the help pages.

Bibliography

Anderson, C. and R. Loynes (1987). *The Teaching of Practical Statistics*. New York: Wiley.

Anderson, R. and T. Bancroft (1952). *Statistical Theory in Research*. New York: McGraw Hill.

Andrews, D. and A. Herzberg (1985). *Data: A Collection of Problems from Many Fields for the Student and Research Worker*. New York: Springer-Verlag.

Becker, R., J. Chambers, and A. Wilks (1998). *The New S Language: A Programing Environment for Data Analysis and Graphics* (revised ed.). Boca Raton, FL: CRC Press.

Belsley, D., E. Kuh, and R. Welsch (1980). *Regression Diagnostics: Identifying Influential Data and Sources of Collinearity*. New York: Wiley.

Berkson, J. (1950). Are there two regressions? *Journal of the American Statistical Association 45*, 165–180.

Box, G., S. Bisgaard, and C. Fung (1988). An explanation and critique of Taguchi's contributions to quality engineering. *Quality and Reliability Engineering International 4*, 123–131.

Box, G., W. Hunter, and J. Hunter (1978). *Statistics for Experimenters*. New York: Wiley.

Carroll, R. and D. Ruppert (1988). *Transformation and Weighting in Regression*. London: Chapman & Hall.

Chambers, J. and T. Hastie (1991). *Statistical Models in S*. London: Chapman & Hall.

Chatfield, C. (1995). Model uncertainty, data mining and statistical inference. *JRSS-A 158*, 419–466.

Cook, J. and L. Stefanski (1994). Simulation-extrapolation estimation in parametric measurement error models. *Journal of the American Statistical Association 89*, 1314–1328.

Dahl, G. and E. Moretti (2003). The demand for sons: Evidence from divorce, fertility and shotgun marriage. Working paper.

Dalgaard, P. (2002). *Introductory Statistics with R*. New York: Springer.

Davies, O. (1954). *The Design and Analysis of Industrial Experiments*. New York: Wiley.

de Boor, C. (2002). *A Practical Guide to Splines*. New York: Springer.

de Jong, S. (1993). Simpls: An alternative approach to partial least squares regression. *Chemometrics and Intelligent Laboratory Systems 18*, 251–263.

Draper, D. (1995). Assessment and propagation of model uncertainty. *JRSS-B 57*, 45–97.

Draper, N. and H. Smith (1998). *Applied Regression Analysis* (3rd ed.). New York: Wiley.

Efron, B. and R. Tibshirani (1993). *An Introduction to the Bootstrap*. London: Chapman & Hall.

Faraway, J. (1992). On the cost of data analysis. *Journal of Computational and Graphical Statistics 1*, 215–231.

Faraway, J. (1994). Order of actions in regression analysis. In P. Cheeseman and W. Oldford (Eds.), *Selecting Models from Data: Artificial Intelligence and Statistics IV*, pp. 403–411. New York: Springer-Verlag.

Fisher, R. (1936). Has Mendel's work been rediscovered? *Annals of Science 1*, 115–137.

Fox, J. (2002). *An R and S-Plus Companion to Applied Regression*. Thousand Oaks, CA: Sage Publications.

Frank, I. and J. Friedman (1993). A statistical view of some chemometrics tools. *Technometrics 35*, 109–135.

Freedman, D. and D. Lane (1983). A nonstochastic interpretation of reported significance levels. *Journal of Business and Economic Statistics 1*(4), 292–298.

Garthwaite, P. (1994). An interpretation of partial least squares. *Journal of the American Statistical Association 89*, 122–127.

Hamada, M. and J. Wu (2000). *Experiments: Planning, Analysis, and Parameter Design Optimization*. New York: Wiley.

Hastie, T., R. Tibshirani, and J. Friedman (2001). *The Elements of Statistical Learning: Data Mining, Inference, and Prediction*. New York: Springer.

Hsu, J. (1996). *Multiple Comparisons Procedures: Theory and Methods*. London: Chapman & Hall.

Ihaka, R. and R. Gentleman (1996). R: A language for data analysis and graphics. *Journal of Computational and Graphical Statistics 5*(3), 299–314.

John, P. (1971). *Statistical Design and Analysis of Experiments*. New York: Macmillan.

Johnson, M. and P. Raven (1973). Species number and endemism: the Galápagos Archipelago revisited. *Science 179*, 893–895.

Johnson, R. and D. Wichern (2002). *Applied Multivariate Statistical Analysis* (5th ed.). New Jersey: Prentice Hall.

Lentner, M. and T. Bishop (1986). *Experimental Design and Analysis*. Blacksburg, VA: Valley Book Company.

Little, R. and D. Rubin (2002). *Statistical Analysis with Missing Data* (2nd ed.). New York: Wiley.

Longley, J. (1967). An appraisal of least-squares programs from the point of view of the user. *Journal of the American Statistical Association 62*, 819–841.

Maindonald, J. and J. Braun (2003). *Data Analysis and Graphics Using R*. Cambridge: Cambridge University Press.

Mazumdar, S. and S. Hoa (1995). Application of a Taguchi method for process enhancement of an online consolidation technique. *Composites 26*, 669–673.

Morris, R. and E. Watson (1998). A comparison of the techniques used to evaluate the measurement process. *Quality Engineering 11*, 213–219.

Partridge, L. and M. Farquhar (1981). Sexual activity and the lifespan of male fruitflies. *Nature 294*, 580–581.

Pinheiro, J. and D. Bates (2000). *Mixed-Effects Models in S and S-PLUS*. New York: Springer.

R Development Core Team (2003). *R: A language and environment for statistical computing*. Vienna, Austria: R Foundation for Statistical Computing. http://www.R-project.org.

Ripley, B. (1996). *Pattern Recognition and Neural Networks*. Cambridge: Cambridge University Press.

Ripley, B. and W. Venables (2000). *S Programming*. New York: Springer-Verlag.

Rodriguez, N., S. Ryan, H. V. Kemp, and D. Foy (1997). Post-traumatic stress disorder in adult female survivors of childhood sexual abuse: A comparison study. *Journal of Consulting and Clinical Pyschology 65*, 53–59.

Rousseeuw, P. and A. Leroy (1987). *Robust Regression and Outlier Detection*. New York: Wiley.

Schafer, J. (1997). *Analysis of Incomplete Multivariate Data*. London: Chapman & Hall.

Sen, A. and M. Srivastava (1990). *Regression Analysis: Theory, Methods and Applications*. New York: Springer-Verlag.

Simonoff, J. (1996). *Smoothing Methods in Statistics*. New York: Springer.

Stigler, S. (1986). *The History of Statistics*. Cambridge, MA: Belknap Press.

Stolarski, R., A. Krueger, M. Schoeberl, R. McPeters, P. Newman, and J. Alpert (1986). *Nimbus 7* satellite measurements of the springtime antarctic ozone decrease. *Nature* (322), 808–811.

Thisted, R. (1988). *Elements of Statistical Computing*. New York: Chapman & Hall.

Thodberg, H. H. (1993). Ace of Bayes: Application of neural networks with pruning. Technical Report 1132E, Maglegaardvej 2, DK-4000 Roskilde, Danmark.

Venables, W. and B. Ripley (2002). *Modern Applied Statistics with S-PLUS* (4th ed.). New York: Springer.

Weisberg, S. (1985). *Applied Linear Regression* (2nd ed.). New York: Wiley.

Wold, S., A. Ruhe, H. Wold, and W. Dunn (1984). The collinearity problem in linear regression: The partial least squares (pls) approach to generalized inverses. *SIAM Journal on Scientific and Statistical Computing 5*, 735–743.

Index